China, Russia, and the Coming Cool War

Nonproliferation Policy Education Center
Arlington, VA 22209
www.npolicy.org

Printed in the United States of America.

Cover image: Cyber warfare specialists serving with the Maryland Air National Guard's 175th Cyberspace Operations Group engage in weekend training at Warfield Air National Guard Base in Middle River, Md., June 3, 2017 (credit: U.S. Airforce)

China, Russia, and the Coming Cool War

By

Henry Sokolski

Nonproliferation Policy Education Center

The Nonproliferation Policy Education Center (NPEC), a 501(c)3 nonprofit organization, is a nonpartisan, educational organization founded in 1994 to promote a better understanding of strategic weapons proliferation issues. NPEC educates policymakers, journalists, and university professors about proliferation threats and possible new policies and measures to meet them.

For current publications of the Nonproliferation Policy Education Center, please visit www.npolicy.org.

Contents

China, Russia, and the Coming Cool War

Acknowledgments

This research was supported by generous grants from the Sarah Scaife Foundation, the Carnegie Corporation of New York, and the Smith Richardson Foundation.

Introduction

Three years ago, the media released photos revealing China's construction of hundreds of new intercontinental ballistic missile silos and Russia was preparing its brutal invasion of Ukraine. Now, even the most ardent of arms controllers find themselves resigned to the emergence of a major strategic contest pitting Beijing and Moscow against the West.

This book aims to clarify the character of this rivalry and how the United States and its allies might prevail without suffering a total war or a nuclear exchange.

The volume contains four chapters and an appendix. The first chapter, "Winning the Coming Cool War," describes what America and its closest allies must do over the next two decades to prevail. In this coming cool war, the collection, analysis, sharing, and the securing of information will be far more important to secure victory than credibly threatening massive destruction with nuclear or conventional weapons.

The volume's second chapter, "Xi and Putin Are Building More Nukes: How to Compete," focuses on the nuclear challenge China and Russia will pose over the next ten to twenty years and what diplomatic and military steps America and its allies should take to counter it.

The third chapter, "What a Missile-Driven Competition with China Will Look Like," maps out what offensive missilery China and the United States are likely to deploy against one another over the next 20 years.

The volume's fourth and final chapter, "Can Self-Government Survive the Next Convulsion?" clarifies how American domestic affairs could adapt to help fend off emerging Russian and Chinese strategic threats. A key suggestion here is to disperse America's key political, corporate, demographic, industrial, and military assets geographically and to provide much more secure communications.

The volume's final section is an appendix consisting of the introductory chapter of a three-year study NPEC released back in 2021. "China's Civil Nuclear Sector: Plowshares to Swords?" was co-authored with Thomas Cochran. The Pentagon has cited this study in each of its annual Chinese military power assessments since 2021.

The Pentagon has used it to help explain how China will likely have nearly as many nuclear weapons by the end of this decade as the United States currently has deployed. This projection, perhaps more than any other, caused national security analysts to focus on China as long-term strategic adversary.

Henry Sokolski

November 2024

Winning the Coming Cool War[1]

1. An earlier version of this chapter was published in: Phil Beradelli, editor, *IMAGINE: Winning the New Cold War* (Mountain Lake, Maryland: Legacy Books, 2024), pp. 285-92.

Forty years after the Fall of the Berlin Wall, it's tempting to imagine the passing of Putin's Russia and Xi's China as well as those within their orbit — North Korea, Iran, and Iran's proxies. Such dreams, however, are unlikely to be realized unless the West rededicates itself to liberal self-rule, secures crucial communications, and fortifies its infrastructure, military forces, and population against long-range attacks. All of this will require more than maintaining Western military might. It will be necessary to collect, process, share, and secure both classified and unclassified information on an unprecedented scale.

Civics and Public Service

Any defense the West might muster against its foes will also demand that it understand what it is defending: liberal self-rule. Unfortunately, civic literacy in Western democracies, especially in the United States, is in decline.[2] Without a basic understanding of the foundation of self-government – natural equality – though, the West will struggle to sustain the public support needed to win. Experts have suggested various programs to remedy this. Which of these programs deserves support is open to

---

2. Sarah Shapiro, Catherine Brown, "The State of Civics Education," *American Progress*, February 21, 2018, https://www.americanprogress.org/article/state-civics-education/.

debate.[3] But the clear metric for civic health – robust public service – is not.

Take the cases of Israel, Taiwan, Turkey, South Korea, Lithuania, Denmark, Sweden, Norway, Finland, Latvia, Austria, Greece, Estonia, and Switzerland. All demand some form of compulsory military or public service. Meanwhile, the United States, the West's wealthiest and most powerful nation, eliminated the draft just over a half-century ago.

Today, the all-volunteer military that replaced conscription faces severe recruitment and retention challenges.[4] Although younger Americans are still interested in serving, it is unclear how willing our political leaders are to reinstitute compulsory public service.[5] We might yet see a revival of mandatory service, but only with increased appreciation of the threats China

---

3. Jack Miller Center: *A Campaign to Strengthen American Civic Education.* https://jackmillercenter.org/teach-for-freedom-campaign/
4. Blue Star Families, "Military Families Less Likely to Recommend Service: Blue Star Families 2023 Military Family Lifestyle Survey," *PR Newswire*, March 12, 2024. https://www.prnewswire.com/news-releases/military-families-less-likely-to-recommend-service-blue-star-families-2023-military-family-lifestyle-survey-302086905.html.
5. Steve Cohen, "Americans Support Mandatory National Service," *The Hill,* October 15, 2023. https://thehill.com/opinion/campaign/4253664-americans-support-mandatory-national-service/.

and Russia pose, along with enhanced civics education and more effective national appeals to serve.[6]

Protecting Communications

During the Cold War, Washington offered foreign aid and security assistance to lock in other nations' allegiances. Likewise, pledging to inflict massive destruction against the West's enemies was a major alliance lever. In the coming conflict, however, what will be needed to fortify alliances will be different. Traditional foreign aid and nuclear guarantees will help but be less successful in securing our mutual interests than effective information campaigns, wider intelligence-sharing, and increased high-tech military collaboration. In this regard, the single most important enabling capability will be ensuring the secure communication of vast amounts of data.

To what extent America and its allies can provide such communication is unclear. Currently, the West dedicates its protected communications capabilities almost entirely to achieving specific, traditional military missions. Yet, using these capabilities as soft-power force multipliers could be just as important to winning future conflicts as using them to protect military secrets or applying them to destroy enemy assets. Sharing strong encryption

---

6. USAFacts Team, "Military recruitment is down. Why don't young Americans want to join?" *USA Facts*, November 7, 2023. https://usafacts.org/articles/military-recruitment-is-down/.

capabilities with citizens in hostile states could enable them to communicate safely and organize against their governments.

Securing U.S. and allied domestic, overseas and undersea cable communications (or finding more secure alternatives) also would reduce the threat of adversaries hacking into our critical infrastructure and military systems. This, in turn, could make it far easier for Western allies to trade information and collaborate on the most confidential matters. It also would enable populations to live and work away from urban centers, which are becoming increasingly vulnerable to long-range missile attacks. Post-Covid demand for remote work is already propelling this shift. Government-provided or certified secure information services and possible tax incentives could amplify this trend.

Unfortunately, U.S. demand for privacy has declined as the public's inclination to bare and share all on the Internet has surged. As a result, the coherency of government efforts to protect privacy has suffered.[7] To assure a broad, healthy demand for privacy (rather than narrow calls for consumer protection),[8] politicians and

---

7. Shane McNeil, "Charting a Course for Privacy: The Urgency for a Defined Path by the US Government," *Homeland Security Today,* January 16, 2024. https://www.hstoday.us/featured/charting-a-course-for-privacy-the-urgency-for-a-defined-path-by-the-us-government/.
8. Müge Fazlioglu, "Most consumers want data privacy and will act to defend it," *IAPP*, March 22, 2023.

educators will have to clarify how self-government suffers without a proper balance being struck between the public and private spheres.[9]

Spreading Out

Western governments certainly can, and should, do more to secure communications among themselves, their citizenry, and with industry to help protect information privacy and promote prosperity. These goals, after all, are the key reasons why our government created the U.S. Postal Service.[10] But there is another reason for improving communications security and the survivability of key Western assets – it's prerequisite to any serious defense against future long-range attacks.

We already have seen how precise, long-range munitions threats have threatened the security of Ukraine and the Middle East. This threat also continues to grow in East Asia. Precision drones and cruise and ballistic missiles have been used against Ukraine's electrical grid as well as

---

https://iapp.org/news/a/most-consumers-want-data-privacy-and-will-act-to-defend-it/.

9. Henrik Skaug Sætra, "The ethics of trading privacy for security: The multifaceted effects of privacy on liberty and security," *Technology in Society*, Vol. 68, February 2022, available at https://www.sciencedirect.com/science/article/pii/S016079IX21003298.

10. U.S. Postal Service https://about.usps.com/who/profile/ .

Russia's Black Sea fleet, western Russian military bases and energy plants, Saudi military assets and oil facilities, and Western shipping in the Red Sea. Meanwhile, in East Asia, high-accuracy Chinese and North Korean missiles threaten infrastructure and military bases in South Korea, Japan, and Taiwan, not to mention American and allied bases, aircraft carriers, and surface ships.

The Lower 48 could be next. The Pentagon recently revealed that China is developing intercontinental-range, non-nuclear missiles that could strike the United States.[11] What might they target? Essential fixed military and civilian assets, including bases, command centers, key electrical transformers and generators (both nuclear and nonnuclear), major ports, air-traffic-control systems, financial and political centers, large computer servers, corporate headquarters, and more are all strong possibilities.

If so, what's our defense? First, some of these targets will have to be hardened or actively defended. Second, most will also need to be distributed, multiplied, camouflaged, decoyed, and, where possible, made mobile. Deploying assets in this manner should make an adversary's efforts to eliminate them exponentially more difficult.

---

11. Ken Moriyasu, "China expanding ways to target continental U.S., Pentagon report says," *Nikkei Asia*, October 19, 2023, available at https://asia.nikkei.com/Politics/International-relations/US-China-tensions/China-expanding-ways-to-target-continental-U.S.-Pentagon-report-says..

Without secure communications, though, none of these measures will work. At present, military command, control and communications systems can be lobotomized. Remote governmental and private workers are also vulnerable to having their vital information stolen and connections to their employers broken. Result: Without secure communications, decentralizing and distributing government and private offices and workers becomes a national security problem rather than a solution.

If we install secure communications, unpopulated areas (e.g., much of North America) can safely host government and corporate hubs that can function even after their home offices have been knocked out. Moreover, Western countries should have an easier time geographically distributing their government functions than China, Russia, or their proxies. Autocracies fear sharing or dispersing power; self-governing nations welcome it. The West, as a result, should find strength and renewed purpose in wider and more equal geographic distribution of political power and financial wealth.

Meanwhile, the West's energy sector, especially its electrical production, would benefit from more resilient national, distributed, micro, and islanded supply options. Greater attention needs to be paid to fortifying existing electrical and energy-distribution networks so they can support one another in emergencies. America and its allies also need to ensure that their electrical supply systems can operate independently to power critical nodes such as key shipping and airports, military bases and command centers.

Likewise, concentrations of military value should be proliferated, distributed, and made more mobile. The United States is currently spending billions to modernize its intercontinental ballistic missile fleet, yet the Pentagon is basing these missiles in vulnerable, fixed silos. This has prompted some[12] to recommend making them more mobile.[13] In addition, there are growing concerns about the vulnerability of America's strategic command, control, and communications systems.

The fix in all these cases is to make it much more difficult for Russia or China to locate and knock out these assets. This will require multiplying their possible locations, decoying and camouflaging them, and making them mobile. The Pentagon should explore rail, shipping-container, and commercial aircraft basing for these assets. The idea would be to force hostile states to acquire ever more missiles and warheads at greater expense than it will cost the West to proliferate, hide, and defend these assets.

As for Western conventional forces, building ever larger, manned multi-mission planes, vehicles, and ships with ever more capable, long-range standoff systems is no

---

12. "China's Emergence as a Second Nuclear Peer," The Center for Global Security Research at Lawrence Livermore Laboratory, Spring 2023, available at https://cgsr.llnl.gov/content/assets/docs/CGSR_Two_Peer_230314.pdf.
13. "America's Strategic Posture," Congressional Commission, October 2023, available at https://armedservices.house.gov/sites/republicans.armedservices.house.gov/files/Strategic-Posture-Committee-Report-Final.pdf.

longer financially or logistically sustainable. Instead, the United States and its allies must design specific packages – such as short-range, unmanned surface and underwater vessels, missiles, mines, and drones – tailored to block hostile states from attacking key military, communications, and infrastructure nodes.[14] Doing so will require reforming our military acquisition, classification, production policies, and organization to catalyze military innovation and increased allied production and development collaboration.

Finally, at both the tactical and strategic level, the West will have to deemphasize knocking out all of its enemies' military systems in early, prompt strikes and focus instead on making its own military forces more difficult to knock out. Adopting such a posture should make effective first strikes against us (no matter how many missiles or warheads our adversaries might acquire) all but impossible.

The Space Card

It's been nearly a century since airpower theory – the idea that threatening to decimate concentrations of political, demographic, industrial, and military capital with aerial

---

14. Bryan Clark and Dan Patt, "Hedging Bets: Rethinking Force Design for a Post-Dominance Era," Hudson Institute, February 20, 2024. https://www.hudson.org/defense-strategy/hedging-bets-rethinking-force-design-post-dominance-era-bryan-clark-dan-patt.

strikes can quickly win or deter wars – became a dominant military organizing principle for the world's most powerful nations.[15] In a desire to win such wars, nuclear-armed states gave the ultimate authority to launch devastating nuclear strikes to their heads of government (e.g., the president, prime minister, or supreme leader). They also enabled these commanders-in-chief to operate under unprecedented levels of secrecy to inflict massive destruction instantaneously if ever deemed necessary. But recently, another theory of armed conflict — new generation warfare — has emerged. In contrast to airpower theory, this strategic model holds the promise of disabling enemies without physically obliterating them. What's required to do this is advanced processing, dissemination, and protection of unprecedented amounts of data.

Unmanned, precision offensive and defensive weapons will depend on artificial intelligence-fortified target acquisition and high-fidelity reconnaissance and surveillance transmissions. Advanced computing, directed energy, electronic warfare, and optic and laser communication technologies also will play major roles in the transition to this new form of warfare. Finally, the acquisition and transmission of imagery, communications, and data will depend heavily on space-based systems. By achieving space superiority and committing to the

---

15. Henry Sokolski, "Hiroshima at 75: A Peak into Our nuclear Future," *The American Interest*, August 6, 2020. https://npolicy.org/hiroshima-at-75-a-peek-into-our-nuclear-future-the-american-interest/

collaboration and sharing of sensitive information, the West will gain a better chance of staving off and, if necessary, defeating greater naval, air, and land forces, nuclear or nonnuclear.[16]

Fortunately, we currently enjoy a comparative advantage in the development of reusable, heavy-lift space launch vehicles; large internet and communications space constellations; and space imagery, signals, and data processing. If we redouble these efforts, America and its spacefaring allies should be able to reconstitute and refuel existing and future space-based systems, making them more resilient and maneuverable (i.e., capable of moving much more freely and quickly from one orbit to another).[17] This, in turn, should give the West superior passive and active space defenses and commercial space systems.

Victory?

Can we surmount these challenges? Absolutely. Will we? That's unclear. It's certainly unlikely if we believe China, Russia, and their proxies pose little or no threat to liberal

---

16. "Over-classification: How Bad Is It, What's the Fix?" Nonproliferation Policy Education Center, March 28, 2023. https://npolicy.org/over-classification-how-bad-is-it-whats-the-fix-occasional-paper-2303/.
17. Brent Ziarnick, "Space Power Theory and Space Deterrence," February 9, 2023. https://npolicy.org/wp-content/uploads/2023/02/20230209-Ziarnick-Spacepower-Theory-and-Deterrence-1.pdf.

democracy. Nor is the West likely to prevail if it thinks it can win future conflicts by simply mustering ever more arms or threatening massive destruction.

Those were Cold War competitive levers. What's coming is different: keeping key military, political and commercial functions immune to being knocked out, and communicating, protecting, and manipulating essential information will now be crucial.

Xi and Putin Are Building More Nukes: How to Compete[1]

1. An earlier version of this chapter was published in: *Nuclear Danger & the NPT* (Arlington, VA: Nonproliferation Policy Education Center, 2024), pp. 122-33, available at https://npolicy.org/wp-content/uploads/2024/06/Book-Nuclear-Danger-and-the-NPT.pdf.

Three years after *The Washington Post* first released photos of China's construction of more than 300 nuclear missile silos and the Pentagon forecasted China would have at least 1,000 nuclear weapons by 2030, our government is still pondering what it should do. Meanwhile, Russia is developing entirely new nuclear weapons systems — hypersonic nuclear-armed missiles, nuclear anti-satellite weapons, and long-range nuclear torpedoes.

The Pentagon's response to date is to modernize America's nuclear bombers, submarines, and missiles. This modernization, however, won't be complete for another decade.

It also won't be enough. Matching China and Russia's nuclear expansion goals missile-for-nuclear-missile, warhead-for-nuclear-warhead, much less clearly staying ahead, could require a tripling or more of America's current nuclear deployments of 1,770 nuclear weapons.

That's a tall order. America's nuclear weapons pit production, nuclear submarine, and intercontinental ballistic missile modernization programs have repeatedly run over budget and continue to slip behind schedule.[2]

---

2. Stephen Young, "Another reason to cancel the Sentinel missile: the rising cost of its nuclear warhead," *Bulletin of the Atomic Scientists*, May 23, 2024. https://thebulletin.org/2024/05/another-reason-to-cancel-the-sentinel-missile-the-rising-cost-of-its-nuclear-warhead/.

China, in contrast, has a proven track record of deploying new weapons systems at record pace. As for Russia, it has the world's largest nuclear weapons materials production base.

Our closest allies know this. South Korea, Japan, and Poland want the United States to base nuclear weapons on their soil or to get nuclear weapons of their own. This presents two problems. First, overseas U.S.-based nuclear weapons sites are target sinks. The United States placed thousands of its weapons in Taiwan, Japan, South Korea, and NATO states in the 1950s and 1960s because the Pentagon lacked sufficient long-range systems to deliver them. Now, it does. That's why the Pentagon removed nearly all its forward deployed nuclear weapons in Europe and Asia. Redeploying them would be a military step backwards.

Second, letting our closest allies go nuclear will increase alliance tensions. Historically, this happened when the British, French, Israelis, and Pakistanis went nuclear.[3] Yet, as China and Russia increase their nuclear capabilities, the demand signal for our allies to go nuclear is rising.

What, then, should we do?

---

3. Zachary Keck, *Atomic Friends: How America Deals with Nuclear-Armed Allies*, Rowman and Littlefield, August 2022. https://npolicy.org/atomic-friends-how-america-deals-with-nuclear-armed-allies/.

Adjust America's Nuclear Strategy

America's military has long required its strategic nuclear forces to be able to deliver a devastating counter punch against Russia or China as soon as either launch a nuclear attack against the United States. The aim of such a counterstrike would be to limit any additional damage Russia and China might otherwise inflict on the United States. This puts a premium on launching U.S. nuclear forces as soon as they come under attack — either right after a warning of a Chinese or Russian launch or immediately after Chinese and Russian nuclear weapons detonate on US soil.

These demanding requirements determined much of the quality and quantity of America's current nuclear force. Fifty years ago, the United States had over 31,000 nuclear weapons that were relatively immune from being knocked out. The US military aimed these weapons at a limited number of relatively soft, fixed Russian long-range nuclear systems. This made prioritizing damage limitation eminently sensible.

Accomplishing it now, however, is far more challenging. Many Chinese nuclear missiles are mobile and based in three thousand miles of deep underground tunnels. China is also basing hundreds of its missiles in silos and developing new, stealthy long-range bombers, advanced nuclear ballistic missile submarines, and hypersonic missiles.

Then there's China's ally – Russia. It has thousands of nuclear weapons mounted on mobile intercontinental ballistic missiles launchers, strategic bombers, nuclear submarines, and theater nuclear missiles. Because China's and Russia's nuclear forces are large, relatively invulnerable, and growing, continuing to prioritize the mission of limiting damage could easily require doubling or tripling America's current deployed nuclear force of roughly 1,770 weapons.

That would be difficult. First, the Pentagon and Energy Department have struggled to update the current U.S. force of intercontinental ballistic missiles, nuclear warheads, ballistic missile submarines, and bombers. All these programs are significantly behind schedule or over budget. The prospects of the United States quickly outproducing China and having confidence in its ability to destroy remaining Russian and Chinese nuclear systems after a first strike are slim and dwindling. In the 1960s we had a clear shot; now it's unlikely.

Worse, the key instrument to achieve this damage limitation mission — America's intercontinental ballistic missile force— is itself vulnerable as this system is based in relatively soft, fixed silos. It is unclear how many of these siloed missiles would survive a first strike unless they were launched on warning or under attack. As the recent Congressional Commission on the Strategic Posture of the United States noted, however, launching under these circumstances is hardly desirable.

This suggests adjusting America's nuclear strategy. The Pentagon should still try to limit damage after the United States is attacked. But it should also recognize that prioritizing damage limitation will become more expensive and difficult as the numbers of Russian and Chinese nuclear systems, their location uncertainty, and their known positions increase.

Rather than significantly expanding the U.S. nuclear arsenal to achieve a high level of damage limitation, the United States should make it far more difficult for China or Russia to disable or destroy U.S. and allied nuclear forces. This recommends transitioning all its siloed intercontinental ballistic missiles to mobile platforms and camouflaging and decoying them along with U.S. strategic submarines and bombers. Alternative basing on commercial planes, trains, and ships also should be explored. For key fixed targets, active defenses and hardening may be necessary.

These efforts would exponentially increase the number of possible targets China and Russia would have to strike to eliminate America's strategic nuclear forces. Instead of Moscow and Beijing needing hundreds of nuclear-armed missiles and thousands of nuclear warheads to accomplish this, they would need tens of thousands of warheads — and still not be confident of knocking out all of America's nuclear force.[4] The daunting nature of this requirement

4. The Pentagon believes China is developing intercontinental nonnuclear missiles and may deploy them by 2030. If so, infrastructure nodes (e.g., electrical

should assure future American presidents that they would never have to threaten to "use or lose" America's nuclear weapons to deter Moscow or Beijing.

Unfortunately, strategic deterrence requires more than nuclear weapons. As Israel discovered with Iran's massive April 2024 missile strike, nuclear arms don't always deter enemy states from launching nonnuclear attacks, even ones that have strategic consequences.

Also, no nuclear force, no matter how well designed, is guaranteed to neutralize or deter all nuclear threats. As Putin has demonstrated in his war against Ukraine, nuclear-armed states can use nuclear threats to deter nuclear-armed alliances (e.g., NATO) from tipping the military scales against them in nonnuclear wars. With any bad luck, China, Russia, or North Korea may repeat Putin's nuclear dares over Ukraine to stare down the West's will to arm its allies and defend itself against conventional attacks.

In these cases, the proper remedy is more nonnuclear than not. To deter conventional Russian, Chinese, or North Korean attacks, the United States and its allies will need

---

generating plants and transformers, key air and seaports, bridges, communication systems, large computer servers for major companies and financial exchanges, natural gas and oil transmission choke points, etc.) may be targeted. This would recommend proliferating or distributing as much of these assets as possible or hardening or actively defending them.

sufficient conventional forces to destroy invading armor, ships, and planes. Increasingly, these forces will consist of unmanned weapons fed with copious amounts of targeting data and fortified with artificial intelligence. This, in turn, will require air and space-based target acquisition and navigation systems to convey vast amounts of information via super secure space, air, and seabed-based broad bandwidth communications networks. The United States will also need to harden and protect these secure networks against attack.

Leverage U.S.-Allied Nuclear Diplomacy

This brings us to the second required strategic adjustment: Employing new forms of nuclear diplomacy to counter Russian and Chinese nuclear threats and complaints against U.S. nuclear sharing geared to drive a wedge between America and its allies.

The conventional wisdom is that arms control and nonproliferation diplomacy are impossible without the cooperation and engagement of Moscow and Beijing. Since Russia and China currently see America and its allies as their enemies, U.S. and allied diplomats cannot expect to reach any major, new arms control or nonproliferation treaty agreements. This has discouraged American and allied diplomats from proposing any bold arms control or nonproliferation initiatives.

Meanwhile, Russia and China have been busy. Russia has proposed "demilitarizing" space, threatening to use its theater nuclear weapons if NATO helps Ukraine directly, and demanding that NATO remove all American nuclear weapons from Europe. Not to be outdone, China has promoted agreements on no first use of nuclear weapons and warned Washington against supplying nuclear submarines to Australia or redeploying nuclear weapons on South Korean soil. China also has refused to engage in "good faith" negotiations to limit or control nuclear arms as required by Article IV of the Nuclear Nonproliferation Treaty (NPT) until China reaches nuclear parity with the United States. Similarly, Russia has frozen arms control talks with the United States until NATO stops supporting Ukraine.

Xi's and Putin's aim is to weaken America and drive a wedge between it and its allies. American and allied diplomats, instead, are mostly focused on creating new confidence-building measures, arms control understandings, and binding treaties with Beijing and Moscow. As desirable as securing such understandings may be, leveraging the West's comparative military and technical advantages against China and Russia is more important.

Recently, the Biden Administration proved this point. It asked Japan to demand that Russia back a United Nations Security Council resolution affirming Russia's Outer Space Treaty obligation not to put nuclear weapons into orbit. The resolution had the support of 62 nations, most of whom rely upon U.S. and Western satellites to meet their communications and imagery needs. Russia and China tried to turn this resolution to their advantage by demanding all weapons in space be banned, including ones permitted by the treaty. It didn't work: Russia, caught testing a satellite designed to carry a nuclear weapon, was isolated. It alone voted against the Japanese resolution. China, not wanting to be on the losing side, abstained.

The United States and its allies need to double down on such diplomacy. An ideal way to do so is to focus on nuclear sharing. Late in 2023, the Pentagon spotlighted[5] China's probable use of its "peaceful" fast reactor and plutonium recycling programs to match or exceed America's nuclear force by 2035.[6] At the same time, China has repeatedly warned Washington not to share U.S. nuclear weapons with South Korea or Japan.[7]

---

5. Yuki Kobayashi, "China's Fast Breeder Reactor Operating? Possibility of Accelerating Nuclear Arms Race," *SPF China Observer*, November 30, 2023. https://www.spf.org/spf-china-observer/en/eisei/eisei-detail006.html.
6. Idrees Ali and Phil Stewart, "China likely to have 1,500 nuclear warheads by 2035: Pentagon," *Reuters*, November 29, 2022. https://www.reuters.com/world/china-likely-have-1500-nuclear-warheads-by-2035-pentagon-2022-11-29/.
7. Henry Sokolski, "Let's call China's bluff on nukes and plutonium," *The Hill*, August 23, 2023. https://thehill.com/opinion/national-security/4164909-lets-call-chinas-bluff-on-nukes-and-plutonium/.

Presumably, Beijing does not want Washington to undermine China's nuclear build up by redeploying U.S. nuclear arms to its Asian neighbors.

This hypocrisy presents an opportunity.

From a military scientific perspective, Washington has little to gain from redeploying its nuclear arms to Asia. It originally based U.S. nuclear weapons overseas when it lacked intercontinental nuclear missiles, long-range air and sea-launched cruise missiles, and a reliable intercontinental-range bomber force. By the late 1980s, it had all these modern nuclear delivery systems and no longer had to risk placing its nuclear arms in vulnerable bases overseas. As a result, Washington dramatically reduced the number of weapons it based in NATO countries (from 7,000 to roughly 100) and in East Asia (from more than 3,000 to zero).[8]

South Korea's president, Yoon Suk Yeol,[9] and former Japanese president, Shinzo Abe,[10] have spoken of the

---

8. Robert Norris, William Arkin, and William Burr, "Where they were," *The Bulletin of Atomic Scientists*, November/December 1999. https://www.archives.gov/files/declassification/pidb/meetings/where-they-were.pdf.
9. Chloe Sang-Hun, "In a First, South Korea Declares Nuclear Weapons a Policy Option," *The New York Times*, January 12, 2023. https://www.nytimes.com/2023/01/12/world/asia/south-korea-nuclear-weapons.html.

possible need for the United States to redeploy nuclear weapons to Japan and South Korea. Washington, so far, has demurred and instead offered increased nuclear planning consultations and exercises. For the moment, this has satisfied South Korea and Japan.

It's unclear how long this will work. As Chinese and North Korean nuclear arsenals grow, it is likely Japanese and South Korean interest in acquiring nuclear arms will as well. To satisfy this demand, the RAND Corporation has suggested an intermediate fix: Have Seoul pay to construct nuclear storage sites in South Korea and modernize 100 American tactical nuclear weapons that Washington otherwise plans to dismantle.[11] Washington could then keep these refurbished weapons in the United States for quick deployment if South Korea should ever need them.

Validating this proposal's value, China has already objected to it.[12] This opens a door: Washington should tell

---

10. Jesse Johnson, "Japan should consider hosting U.S. nuclear weapons, Abe says," *The Japan Times*, February 27, 2022. https://www.japantimes.co.jp/news/2022/02/27/national/politics-diplomacy/shinzo-abe-japan-nuclear-weapons-taiwan/.
11. "Options for Strengthening ROK Nuclear Assurance," RAND Corporation, October 29, 2023. https://www.rand.org/pubs/research_reports/RRA2612-1.html.
12. Park Chan-kyong, "Should South Korea 'scare Kim' with US nuclear bombs? 'China and Russia would raise

Beijing it has no desire to redeploy nuclear weapons either to South Korea or Japan, but that China's and North Korea's continued nuclear build up produces a demand signal for Seoul and Japan. If Beijing is serious about preventing future U.S. nuclear deployments to the region, China need only reduce this signal.

A simple way to do this would be for China to freeze further construction and operation of plutonium recycling and fast breeder reactor plants, which can produce hundreds of bombs worth of super-weapon grade plutonium a year. Compliance could be verified remotely from space. In exchange, the United States might offer to place a moratorium on any forward deployment of its nuclear weapons to East Asia. This moratorium would last as long as China kept the operation and construction of its "peaceful" plutonium recycling and fast reactor activities frozen. Washington could go further and offer Beijing a regional freeze on South Korean or Japanese fast reactor and plutonium recycling activities and even promise to freeze commercial deployment of American fast reactors.

In Europe, nuclear sharing presents different challenges. In 2023, Putin redeployed Russian nuclear weapons to Belarus. Since then, he has protested American plans to refurbish nuclear storage sites in the United Kingdom and

---

hell'" *South China Morning Post,* May 20, 2024. https://www.scmp.com/week-asia/politics/article/3263120/should-south-korea-scare-kim-us-nuclear-bombs-china-and-russia-would-raise-hell.

possibly to redeploy nuclear weapons there. Russian officials have also protested Sweden's willingness to station U.S. nuclear weapons in wartime and Polish pleas for Washington to deploy nuclear weapons in Poland.[13] Finally, Russia has demanded the United States remove all its nuclear weapons from Europe and suggested that Russia might consider removing its weapons from Belarus only if the United States and NATO "refrained from undermining the security and sovereignty of Russia and Belarus."

The inconsistency of Russia objecting to additional U.S. nuclear weapons deployments to Europe while redeploying Russian nuclear weapons to Belarus is something even the Chinese have spotlighted.[14] Washington should do the same. It should tell Moscow that, while the United States would prefer to take its nuclear weapons out of Europe, it can't so long as Russia continues to deploy nuclear weapons there. To cut this knot, Washington should offer to remove the nuclear

---

13. John Tirpak, "President Says Poland 'Ready' to Join NATO Nuclear Sharing," *Air and Space Forces Magazine*, April 22, 2024. https://www.airandspaceforces.com/president-says-poland-ready-to-join-nato-nuclear-sharing/.
14. "Remarks by Ambassador Geng Shuang at the UN Security Council Briefing on Threats to International Peace and Security," Permanent Mission of the People's Republic of China to the UN, March 31, 2023. http://un.china-mission.gov.cn/eng/hyyfy/202304/t20230401_11052883.htm.

weapons it has deployed in Europe if Russia redeploys all its theater weapons to bases located east of the Urals and allows effective, mutual inspections.

Meanwhile, the United States and NATO should do everything it can to strengthen its nuclear capabilities in Europe short of forward deploying additional U.S. nuclear weapons there. This would include building new and refurbishing old nuclear weapons storage sites in Europe, selling new nuclear-capable planes (F-35s) to any NATO member that wants them, and creating nuclear escrow accounts for Poland and any other NATO member willing to pay to have the United States refurbish and store on American soil the tactical nuclear weapons the Pentagon otherwise is planning to dismantle.

Exploit the NPT

Where might the United States best promote these ideas? The very place Russia and China have put the most nuclear diplomatic pressure on the United States and its closest allies — the NPT preparatory review conferences leading up to the next NPT Review Conference in 2026.

Reviewing the NPT affords two diplomatic handles. The first is Article IV of the treaty. It requires NPT members to negotiate in good faith measures to end the "nuclear arms race" — something China has refused to do. Instead, China has insisted that it will only negotiate once it has achieved nuclear parity with the United States. This flies in the face of the clear letter and intent of the treaty.[15] The United States and like-minded nations should drive this home while offering Beijing ways to make good on their NPT obligations.

The second handle is Articles I and II of the NPT, which prohibit the transfer of "control" over nuclear weapons to any state that lacks them. The question here is what does "control" mean? China and many developing nations have long questioned the validity and authority of the agreement reached by the United States and the Soviet Union in the late 1960s. As they point out, Moscow and Washington failed to share their mutual understanding with most of the original signatories to the NPT until after many nations signed, and that, in any case, nuclear sharing violates the treaty's spirit.

Allowing their objections without prejudicing our allies' security, Washington should first stigmatize Putin's Belarusian nuclear weapons deployment. Washington should then offer to withdraw US weapons from NATO states if Russia verifiably withdraws its theater weapons locations East of the Urals. Washington could, then, further burnish its nonproliferation credentials by promising not to redeploy American nuclear weapons to East Asia if China suspends its fast-reactor and plutonium recycling-related activities.

---

15. Henry Sokolski, Thomas Grant, "China: A nuclear nonproliferation deadbeat," *The Hill*, August 30, 2022. https://thehill.com/opinion/national-security/3621641-china-a-nuclear-nonproliferation-deadbeat/.

Be Patient

Would China and Russia immediately agree to these nuclear control proposals? No. Eventually, however, they might. After all, at the height of the Cold War, the United States and the Soviet Union struck agreements to demilitarize Antarctica, ban nuclear weapons testing above ground, prohibit weapons of mass destruction in space, and limit the spread of nuclear weapons.

How did this happen? It helped that Stalin had died, and that Khrushchev came into power. It was even more helpful that several major nuclear crises spooked both sides back to the negotiating table and that, by the late-1960s, both knew their strategic forces were becoming immune to being knocked out in a first strike.

These past prerequisites may come into play again. Putin's or Xi's grip on power will eventually give way. America now is committed to making itself less vulnerable to strategic attacks. As for the prospects of a major war, they are increasing. With any luck, these developments will produce the very tensions needed to catalyze sound diplomacy. America should get ready.

What Missile-driven Competition with China Will Look Like[1]

1. An earlier version of this chapter was originally published in *American Purpose* October 20, 2021, available at https://www.persuasion.community/p/missile-driven-competition-china-sokolski .

In 2026 New START will run out. Both Biden and Trump are on record wanting to draw China into nuclear arms negotiations. Assuming that was possible, what would that look like?

The short answer is unclear. The long answer is whatever deal might be struck must reflect the long-term missile-driven military competition that's already underway with Beijing. This arms rivalry will create ample opportunities for diplomacy as well as for increased military risks and miscalculation. It will come in several stages, most likely along the following lines.

Phase I: 2020-2030

During the early years of the first phase of military missile-driven competition with China, from approximately 2020 to 2030, the United States will be able to strike China's homeland with many more long-range missiles than China could use against the continental United States. Initially, all these intercontinental-range missiles, on both sides, will be nuclear-armed. By 2030, however, some of the missiles will be nonnuclear.

From now through the mid-2030s, China will have far more nonnuclear short, medium, and intermediate range missiles and drones in East Asia than the United States or its allies will. By 2030, though, the United States and its East Asian allies will have begun closing this gap with some number of short, medium, and intermediate range nonnuclear missile systems based in South Korea, Japan, U.S. territories, and perhaps Taiwan, Vietnam, and the Philippines. In addition, both China and the United States will begin deploying advanced bombers capable of delivering large numbers ("swarms") of nonnuclear missiles and drones.

At the end of this first phase, the United States and its allies could seek to put China on the defensive by targeting assets essential to the Chinese Communist Party's (CCP's) rule over provinces that would rather go their own way. One might hope that this type of targeting would encourage the CCP to invest much more in active and passive defenses on the mainland to protect targets that are currently undefended. These CCP defensive investments would come at the expense of China's otherwise investing more in offensive strike systems.

U.S. and Chinese active defenses during this phase will continue to be less than perfect but effective enough to force both sides to increase the size of their offensive missile arsenals. The United States will continue to rely very heavily on space-based systems to meet its military command, communication, control, intelligence, and surveillance requirements. China and Russia will rely on them too, but not as heavily. For most of this period, U.S. and allied space-based systems will be vulnerable to Russian and Chinese ground-based lasers. These systems can dazzle and damage allied satellite optical sensors and by the mid to late 2020s will be able to disable the satellites themselves, destroying their metal skins. During this same period, rendezvous satellites, designed to refuel, repair, and reposition spacecraft into their desired orbits, could be used as anti-satellite weapons to de-fuel, damage, and disable them by pulling them out of position.

Phase II: 2025-2035

China will continue to develop accurate nonnuclear missiles that can target the continental United States with maneuverable re-entry vehicles and hypersonic technologies. China would also use intermediate, medium, and shorter-range versions of such missiles to threaten America's regional friends and allies. This, in turn, will increase U.S. and allied interest in developing enhanced passive and active missile defenses, including ultra-performance concrete protected structures, high mobility missile launchers, and advanced decoys and camouflage. It would also increase American investments in low-vulnerability space-based command, control, communication and surveillance systems as well as missile defense and anti-satellite lasers. In general, a premium would be placed on developing a variety of high fire-rate, low cost-per-shot active defenses and using passive defenses to increase the number of missiles that the Chinese would have to fire destroy any given target. Consideration will be given on both sides to space-based missile interceptors.

Against such U.S. and allied efforts, Beijing might hedge its strategic bets by developing a large "peaceful" nuclear fissile material production base, one it could ramp-up relatively quickly to produce large numbers of nuclear weapons, if it chose. China could build such an infrastructure largely by maintaining its current "civilian" nuclear plans.

Phase III: 2030-2040

Once both the United States and China have larger numbers of long- and shorter-range precise strike systems and much more robust active and passive missile and air defenses, the possibility of major missile exchanges would be less attractive to both sides than it might seem today. So, too, would the early use of nuclear weapons. By 2035, the United States should have much more resilient and survivable space-based systems to support its military. Drones and hypersonic weapons could still penetrate missile defenses to hit both stationary and mobile military assets as well as critical infrastructure and political targets.

In this regard, the United States might enjoy some advantage: Unlike China, America's government enjoys far more popular support than the CCP. Also, China would still be far more urbanized than the United States, making it more difficult for China to target all of the political nodes in the United States that might matter.

Diplomatic Opportunities

Given China's regional military advantages and the nuclear superiority of the United States over the next decade, Washington and its East Asian allies would likely have difficulty persuading China to make any arms control deals of the counting-numbers sort previously reached with the Russians. With this in mind, diplomatic proposals should initially focus on new ideas and rules that would be awkward for China to reject and would not turn on relative numbers per se.

It might be possible and useful, for example, to consider negotiations to establish a prohibition against using nuclear weapons against cities. This would make sense militarily and politically even if China did not agree.

First, as noted, China is more urbanized and so more vulnerable to attack than the United States and yet more likely than the United States to threaten to target cities. Also, it would be to the U.S. and allied advantage to put Beijing on the spot to agree not to target cities in Japan and South Korea (two countries that are even more urbanized and vulnerable than China). Finally, pushing such a diplomatic effort would alert China's most urbanized irredentist provinces that in any war, the United States would spare them. Given that the Chinese Communist Party leadership is reluctant to lose these provinces in war, this message would indirectly strengthen America's hand in deterring the Chinese from launching major military operations that might prompt such counterstrikes.

There is a second diplomatic initiative that might help Washington maintain alliance cohesion with Tokyo and Seoul and reduce the prospect of any massive Chinese nuclear ramp-up. The United States, Japan, and South Korea could approach Beijing to establish a moratorium on expanding "peaceful" uranium enrichment capacities and commercial spent fuel recycling plants. The logic of such an understanding – which could be the basis of a political deal – would be to trade Japanese, U.S., and South Korean nuclear production restraint for similar Chinese caution.

This proposition should appeal to China, since Beijing is on record criticizing Japan for Tokyo's reprocessing plans (which include opening a massive plant at Rokkasho in 2027 capable of making 1,500 bombs' worth of nuclear explosive plutonium a year). Beijing would also prefer that South Korea, which has expressed an interest in building and operating large reprocessing and enrichment plants of its own, not to do so. Keeping both Japan and South Korea from developing such options is also in America's interest, as it would encourage both Asian allies to work more closely with Washington than they might otherwise.

Third, and related, Washington might take the lead on proposing measures to deter any further Non-Proliferation Treaty (NPT) withdrawals. Discussing this subject would be timely, given immediate concerns about Iran and other Middle Eastern nations, like Turkey and Saudi Arabia, leaving the NPT and South Korea and Japan possibly doing the same. Washington should take this up at the NPT Review Conference, to be held in early 2026. Here, the United States might work with the United Kingdom and France—two permanent members of the UN Security Council—on a country-neutral UN resolution. (A good place to start would be these recommendations.)[2] Once Washington reached agreement with Paris and the UK, it could bring up the matter with Russia in future arms control discussions; Beijing should also be consulted.

Pushing modest political understandings like these would further these goals:

- Providing some substance to the upcoming NPT Review Conference;
- Testing the arms control sincerity of Russia;
- Opening meaningful nonproliferation arms control talks with China and gauging their willingness to deal with the pressing nuclear concerns of Japan and South Korea.

---

2. Pierre Goldschmidt, "Dealing Preventatively with NPT Withdrawal," Nonproliferation Policy Education Center, January 23, 2020. https://npolicy.org/event_file/Pierre_Goldschmidt_Dealing_Preventively_with_NPT_Withdrawal_140220_1509.pdf?ref=americanpurpose.com.

Fourth, the United States and its Asian and European allies could start now to promote rules of the road for rendezvous satellite operations and ground-based lasers. These rules should be designed to reduce the potential use these systems to launch surprise attacks against U.S. and allied space-based assets critical to conducting nuclear and conventional military operations. The United States and its key NATO and East Asian allies would all have an interest in backing such rules to help clarify what actions might constitute acts of war in space.

As China becomes more dependent on space-based command, control, communication, surveillance and targeting systems, clarifying these matters should be of greater interest to Beijing as well. China may also want to offer civil and commercial space programs for export. Therefore, rejecting such rules could prove awkward, as the rules would be useful not just to maintain safe military satellite operations, but to assure that commercial operations would meet accepted international safety and insurance standards for commercial satellite offerings.

Finally, it might make sense to propose limiting the numbers of hypersonic long-range missiles on both sides. Such limitations might be difficult for China to push aside and could give the United States more time to develop such offensive systems, since neither side is likely to deploy many until later in this decade. After 2030, with both sides more equally armed, the United States might be in a better military position to agree safely on additional missile limits on shorter range hypersonics as well. By then, one would hope that the United States would have built a larger and more advanced missile production base than China's, one that gives America a margin of safety.

Can Self-Government Survive the Next Convulsion?[1]

*A version of this chapter was originally published by *American Purpose,* Can Self Government Survive the Next Convulsion?" June 18, 2021, available at https://www.persuasion.community/p/can-self-government-survive-the-next-convulsion.

As Washington ponders how close it came to a coup on January 6 and the nation unmasks from a pandemic that's left over 600,000 dead, here's a question worth asking: How prepared is our country for the next convulsion?

During the nuclear war-spooked 1950s and 1960s, Washington supplied all kinds of answers, designed to get us through the "day after"—bomb shelters, civil defense, line-of-succession laws, remote bunkers for Washington's political elite. These hedges even had a name: continuity of government, or COG.

With Dwight Eisenhower's heart attack, the killing of John F. Kennedy, the near assassination of Ronald Reagan, and the attacks of September 11, we even had a chance to take some of these plans out for a spin. Security experts and officials viewed such decapitation attempts as "lesser included" security threats the truly Big One – an all-out nuclear attack.

Therein lies a problem: What if the next convulsion isn't a lesser included threat to an atomic apocalypse but something different?

We got a peek at this possibility with the pandemic. Covid-19 took two Congressmen's lives. But what if it had killed a majority of members? Or had just kept Congress from convening? For the suspicious, there's evidence that the Chinese military has given thought to designing diseases to disable us. If it were to do so, could we cope?

A totally different deadly disaster is projected in a fictional video clip titled "Slaughterbots."[2] It portrays precision-guided mini-drone assassinations of Congressmen and other Washington officials. Even the video's critics concede that such high-tech assassinations are plausible.[3]

Their advice: Get over it. But would we? Such assaults would be immediate, visible, and likely to shock the public. But, unlike all-out nuclear attacks, they wouldn't inflict massive destruction: Daily life would continue to function unimpeded for weeks or even months. Rather than physically threatening the entire nation, these strikes would target political and economic power centered in Washington, New York, Los Angeles, and the Bay Area. The goal would be not to decimate the United States, but to disable it.

The Benefits of Dispersal

Worried? What about distributing financial and political power away from the coasts and DC? Well, Harry Truman, Eisenhower, Kennedy, and others tried. They

---

1Stop Autonomous Weapons, "Slaughterbots," YouTube, 2018.
https://www.youtube.com/watch?v=9CO6M2HsoIA
3. Paul Scharre, "Why You Shouldn't Fear 'Slaughterbots,'" IEEE Spectrum, December 22, 2017. https://spectrum.ieee.org/why-you-shouldnt-fear-slaughterbots?ref=americanpurpose.com.

had ambitious plans.[4] There were multiple remote presidential bunkers built, efforts to create field offices for all of Washington's major agencies, and remote stockpiling of critical goods, not to mention the development of super-secure emergency communications systems.

With the end of the Cold War, though, our nuclear fears receded—and, with them, our efforts to "prevail" over the threat once at the top of our minds. Even after 9/11, continuity of government concerns, focused mainly on presidential succession, and were relatively short-lived. Norman Ornstein of the American Enterprise Institute, who organized a post-9/11 continuity of government commission, theorized that Congressmen simply couldn't get their heads around the prospect of dying.[5]

If so, merely creating another commission or demanding hearings or legislation is unlikely to be enough to arm us against the future. This prospect suggests that we would do better to tackle the problem outside Washington, from the bottom up.

We have collegiate debate contests every year. We could start by focusing next year's topic on how best to assure

---

4. Garrett Graff, "Raven Rock: The Story of the U.S. Government's Secret Plan to Save Itself-While the Rest of Us Die," https://tinyurl.com/cmw5mzss
5. Norman Ornstein, "The plan for governing if multiple leaders are incapacitated: There isn't one." The Washington Post, October 2, 2020. https://tinyurl.com/5n9yprph

self-government's survival against future acts of God, as well as adversaries foreign and domestic. Next, we could have lawyers and state administrators work on the topic at their National Association of Bar Executives and National Governors Association meetings.

How might that help? Governors have the constitutional authority to make temporary appointments to U.S. Senate vacancies. State legislatures have authority to propose and ratify amendments to the U.S. Constitution, including amendments related to emergency successions (e.g., the Twentieth and Twenty-fifth Amendments). Finally, governors and state officials are the people who would best know how to adjust local laws, institutions, and political practices to preempt popular demands for "temporary" extralegal martial rule in the case of a major hit on Washington.

Encouraging adjustments like these should poll well with the public, which has grown distrustful of Washington and its increasingly cozy relationship with large corporate players. Trump tried to break up the power of certain federal agencies in Washington by sending their employees out to the regions they serve. The initiative was disrupted by the administration's ambient noise; but we could well try it again, with all of Washington's departments and major regulatory agencies.

The country's virtual office experience with Covid-19 actually increased productivity along with reducing life-draining traffic commutes. Congress and the President could make department heads spend more of their time in their field offices than in Washington. This would bring them and their staffs into direct contact with the people they're meant to serve – the local officials and citizens who mostly live far away from Washington and the coasts.

If these moves are combined with antitrust efforts to redistribute the assets of America's largest online and financial giants, the concentration of megacorporate headquarters on the East and West Coasts, along with their workers and wealth, might abate, increasing America's resiliency against possible attacks. These break-ups might also help spread wealth and power to the relatively dispossessed in flyover country.

Testing the Proposition

To signal support for this approach, Presidents can stop giving their State of the Union speeches in person. The current practice concentrates nearly all the federal government's elected officials, top administrators, and Supreme Court justices in a single room, presenting an almost irresistible target. From Jefferson to Wilson, the message was simply sent to Congress and published. We should revive that practice.

The Trump Administration should also scrub the Green New Deal with an eye to strengthening American security. The plan was meant to shift America to a more electrified future, using federal largesse to make a centralized grid system "smarter" and "cleaner." The new system would override both market signals and local deregulatory experimentation through centralized funding, mandates, and subsidies. What's needed is a more distributed electrical system that is more resistant to acts of God (like naturally induced electro-magnetic pulses, bad weather, and earthquakes) and terrorist and foreign attacks (missiles, drones, and cyber hacks). This will require changing regulations and spending.

As for state, county, and municipal government operations, they should get a wire-brushing: They're increasingly part of a growing but shallow administrative state that has us filling out ever more forms to secure permits, licenses, waivers, and other state and federal cost-shared dispensations—but doesn't solve problems.

If federal or state governments stopped functioning, how well would county and municipal governments work? We should experiment to find out. Unplug the computers for a week. Randomly hold county and municipal board meetings outside city and county meeting halls. Conduct unannounced board meetings without the presence of managerial staff.

Is this an argument for the federal government's simply letting go? No. It would make sense to assert more federal control over the chaotic patchwork of state public health service systems that did such a mixed job of tracing and tracking the national coronavirus pandemic. It would make similar sense to create clear standards for the conduct of federal elections. The objective, though, would be the same as the rationale for breaking up financial and political capital and distributing it geographically: Block and retard assaults and seditious acts that might undo self-government.

As for our national security, the Pentagon, Intelligence Community, and State Department would do well to flee Washington: This migration would make it harder for our enemies to unplug our common defense. It also would put an operational premium on developing something that our country and economy sorely need: much more secure public communications systems, like 5G, quantum computing, and robust encryption systems.

Done right, dispersing our national security leadership away from Washington would also afford us a strategic edge over China and Russia, two states whose leadership is even more concentrated than our own, in a handful of cities. Distributing America's financial and political capital would increase our physical and civic resiliency, making both more difficult to target. The opposite would be the case for China and Russia. Their grip is authoritarian; real decentralization risks eviscerating it.

Everything listed that our government should do here would be in defense of the Constitution, which is itself dedicated to the principle of self-rule. If it is strengthened, America will not only be more likely to survive the next convulsion but also to arrest and reverse the erosion of civic trust that has come to mark the nation's devolution into separate countries, one on the East and West Coasts and another in the heartland in between.

Appendix[1]

How Many Nuclear Warheads China Might Acquire by 2030

Abstract

The U.S. Department of Defense determined last fall that China had a nuclear arsenal in the "low 200s," and that Beijing could "at least double" it by 2030. That estimate, however, failed to factor in China's planned "peaceful" plutonium fast breeder reactor program, which will be capable of producing a significant amount of weapons-grade plutonium. If one includes it and China's nonweaponized military stockpile of weapons-grade plutonium, China conservatively could obtain on the order of 1,270 warheads by 2030 — roughly as many as America currently has deployed on its intercontinental-

---

1. An earlier version of this research was co-authored with Thomas C. Cochran, former senior nuclear scientist at the Natural Resources Defense Council and was published in Henry Sokolski, editor, *China's Civil Nuclear Sector: Plowshares to Swords?* Arlington, VA; Nonproliferation Policy Education Center, 2021), pp. 6-24, available at https://npolicy.org/article_file/2102_Chinas_Civil_Nuclear_Sector.pdf .

range ballistic missile force. China has a variety of additional nuclear materials production options that could enable it to further increase this number by a factor of two or more. Production of weapons-grade plutonium from its "peaceful" fast reactor program, however, is the least burdensome approach to enlarge China's arsenal of advanced, two-stage thermonuclear weapons. Before it completes its fast plutonium reactor plants, Washington should explore with Beijing, Tokyo, and Seoul the merits of agreeing to pause commercializing fast reactor fuel cycles, all of which are exceedingly uneconomic in comparison to other forms of nuclear and nonnuclear electricity production.

I. Overview

The projected number of nuclear weapons China might have by 2030 has grown in the last two years. In 2019, the Director of the Defense Intelligence Agency (DIA) said China might "at least double" its arsenal of nuclear weapons by 2030, which the Pentagon estimated in September of last year to be in the "low 200s."[2] That

---

2. See U.S. Department of Defense, *Military and Security Developments Involving the People's Republic of China 2020* (Washington, DC: Office of the Secretary of Defense, September 1, 2020), available at https://media.defense.gov/2020/Sep/01/2002488689/-1/-

suggested China's arsenal might grow to as many as 500 warheads by 2030 "without new fissile production."[3] Three months later, in December of 2020, though, the Federation of American Scientists (FAS) estimated that China had, not in the low 200s, but 350 nuclear weapons.[4] Finally, in February of 2021, Admiral Charles A. Richard, U.S. Commander of Strategic Command, wrote that the Chinese nuclear arsenal "is expected to double (if not triple or quadruple) over the next decade." More recently,

---

1/1/2020-DOD-CHINA-MILITARY-POWER-REPORT-FINAL.PDF.

3. See, Alex Ward, "China may double its nuclear arsenal in just 10 years. Don't panic.," *Vox*, September 2, 2020, available at https://www.vox.com/2020/9/2/21418252/china-military-report-nuclear-double Hans M. Kristensen, "DIA Estimates For Chinese Nuclear Warheads," *Federation of American Scientist*s, May 31, 2019, available at https://fas.org/blogs/security/2019/05/chinese-nuclear-stockpile/.

4. Hans M. Kristensen and Mark Korda, "Nuclear notebook: Chinese nuclear forces, 2020," *Bulletin of the Atomic Scientists*, December 7, 2020, available at https://thebulletin.org/premium/2020-12/nuclear-notebook-chinese-nuclear-forces-2020/. This estimate is comparable to the estimate of 360 nuclear warheads made by Hui Zhang, "China's nuclear force modernization," in Allison Pytlak and Ray Acheson ed., *Assuring destruction forever: 2020 edition*. June 2020, Reaching Critical Will, a project of the Women's International League for Peace and Freedom, available at https://www.reachingcriticalwill.org/images/documents/Publications/modernization/assuring-destruction-forever-2020.pdf

Admiral Philip Davidson, head of U.S. Indo-Pacific Command, confirmed this projection. [5]

This later projection makes sense if one adds the 350 nuclear weapons, we now believe China has (based on unclassified estimates) to the roughly 480 weapons we believe China could make from the weapon-grade plutonium (WGPu) we believe it has stockpiled but not yet weaponized.[6]

These projections, however, only look at existing "military" fissile stocks. If, in addition, one dials in the roughly 1,440 kgs of WGPu China is capable of producing from the two "civilian" breeder reactors it has under construction, and the 110 kgs of WGPu it could recover by processing blanket material from its small experimental fast breeder reactor, Beijing conservatively could acquire not "between 400 and 500 nuclear

---

5. See, Admiral Charles A. Richard, "Forging 21st-Century Strategic Deterrence," *Naval Institute Proceedings*, February 2021, available at https://www.usni.org/magazines/proceedings/2021/february/forging-21st-century-strategic-deterrence and Caitlin McFall, "US losing military edge in Asia as China looks like it is planning for war: US Indo-Pacific Command chief," *Fox News*, March 10, 2021, available at https://www.foxnews.com/politics/us-losing-military-edge-in-asia-as-china-looks-like-planning-for-war-admiral.amp.

6. Here we assume China has 2.9 ± 0.6 tons of WGPu in weapons and available for weapons and 3 to 4 kilograms (kgs) of WGPu per warhead.

weapons,"[7] but most likely at least 1,270 nuclear warheads by 2030[8] — closing in on or exceeding the roughly 1,300 strategic warheads the United States currently has deployed on its intercontinental ballistic missiles.[9]

EXISTING PRC WGPu STOCKS 2021		
WGPu in kgs	Current number of warheads, assuming 3.5 kgs of WGPu/warhead primary	Additional warheads PRC could make with stockpiled WGPu assuming 3.5 kgs of WGPu/warhead primary
2,900	350	480
ADDITION "PEACEFUL" PRC WGPu STOCKS BY 2030		
WGPu in kgs	Additional warheads, assuming 3.5 kgs of WGPu/warhead primary	Total number of warheads by 2030, assuming 3.5 kgs of WGPu/warhead primary
1,550	440	1,270 (+/-275)

Our projections are conservative. The most contingent assumption is that China will follow through on the construction and planned operation of two additional, large fast breeder reactors and a nuclear fuel reprocessing capacity, which it is building (but has not yet completed)

7. Ward, "China May Double," *Vox.*
8. Our estimate is 1,270 ± 280 warheads. For the benefit of those less familiar with statistics we can express this as between 990 and 1550 warheads.
9. The March 2020 START declaration for the United States shows 1,373 warheads on deployed ICBMs, deployed SLBMs and deployed heavy strategic bombers, where one nuclear warhead is counted for each heavy bomber. There are approximately 56 deployed strategic heavy bombers. See https://www.armscontrol.org/factsheets/Nuclearweapons whohaswhat.

to recover the WGPu from breeder reactors' used radial blanket fuel.

Our projections do not include the additional separated plutonium that might be produced in a second large demonstration reprocessing plant it just revealed it has begun constructing (which may come online before 2030),[10] or the nuclear weapons China might produce using highly enriched uranium (HEU) exclusively. Nor do our projections consider China's ability to double the number of warheads by making composite cores (WGPu and HEU) in the thermonuclear primaries. Our projections excluded these options because they would entail new, untested weapons designs. Nor do our projections consider weapons that China might make from plutonium it could extract from the axial blankets in its planned fast reactors or from plutonium it could generate in its thermal power reactors (both heavy water and light water moderated). Extracting more plutonium from its thermal power reactors would require building additional fuel reprocessing capacity. If one dials in these additional nuclear weapons materials' sources, though, the total number of nuclear warheads China might make by 2030 increases by a factor of at least two or more.

---

10. See Hui Zhang, "China starts construction of a second 200 MT/year reprocessing plant," International Panel on Fissile Materials, March 21, 2021, available at http://fissilematerials.org/blog/2021/03/china_starts_construction.html.

None of this is inevitable. China has only just begun construction of its large, 200 tons per year (t/y) spent reactor fuel reprocessing plant. Nor has it completed work on its two 600 MWe fast breeder reactors. These plants are projected to begin operation in three and five years. More important, it is debatable that it is in China's interest to ramp-up its nuclear arsenal so aggressively. China has actually complained about Japan's ability to make more than 2,500 nuclear weapons from the separated civilian plutonium Japan has on its soil. China also has objected to Japan's plans to start operating its Rokkasho commercial nuclear fuel reprocessing plant in 2023.[11] With a design capacity of 800 t of spent fuel per year and operating at an assumed 75 percent capacity factor, Rokkasho could separate at least 1,200 bombs' worth of plutonium or more a year, assuming 5 kgs of reactor-grade plutonium (RGPu) per warhead.

So far, the United States has quietly persuaded Japan to put off opening this large plant, arguing that its operation would likely upset China as well as South Korea (whose nuclear scientists also want to recycle plutonium). If Beijing proceeds with its reprocessing expansion (in 2025 for the first plant and sometime before 2030 for the second plant) and fast reactor programs (in 2023 and 2026), then it must worry that Japan will proceed with its

---

11. See China Arms Control and Disarmament Association, China Institute of Nuclear Information and Economics, "Study on Japan's Nuclear Materials," September 2015, available at http://fissilematerials.org/library/cacda15.pdf.

reprocessing efforts too. This would result in Japan stockpiling many more thousands of nuclear weapons worth of explosive plutonium and also would likely prompt South Korea to proceed with a plutonium recycling program of its own.

How China would respond is anyone's guess but if it chose to go toe-to-toe with its neighbors, China could easily be seen as a nuclear provocateur by India, Russia, and the United States. How this is China's national security interest is, at best, unclear. Nor is there an economic or environmental case for making or using plutonium-based fuels. Much cheaper uranium is and will be plentiful throughout this century.[12] Nor does recycling ease the management of spent reactor fuel.[13] Nuclear scientists and industry, of course, all want their governments to subsidize large nuclear "commercialization" projects but those using plutonium-based fuel can and should be put off.

---

12. See Bunn, Matthew, Hui Zhang, and Li Kang. "The Cost of Reprocessing in China." Project on Managing the Atom, Belfer Center for Science and International Affairs, Harvard Kennedy School, January 2016, available at https://www.hks.harvard.edu/publications/cost-reprocessing-china.
13. See National Research Council 2008. *Review of DOE's Nuclear Energy Research and Development Program* (Washington, DC: The National Academies Press, 2008), pp. 73-76, available at https://www.nap.edu/cart/download.cgi?record_id=11998&file=73-76.

This, then, brings us to China's surplus uranium enrichment capacity, which is large and could be used to make uranium-fueled bombs or composite plutonium-uranium cores. As noted in the analysis below, China could use only a portion of its unsafeguarded enrichment capacity to make many hundreds of additional single-stage weapons with yields in the scores of kilotons, or higher. Again, there's hardly any economic justification for having so much uranium enrichment capacity. Currently, there is more than enough uranium enrichment capacity within China to meet its projected power reactor requirements. Globally, there is enough to service not just China's, but Japan's, South Korea's, Europe's, and America's commercial needs as well.

What, then, does this study recommend? Three things:

- First, the United States should urge China again to join the United States, as well as South Korea and Japan, in sharing information on their current plutonium and enriched uranium holdings and production capacities. Japan and South Korea already do this. The United States did so in 1996. China did so until 2017, but hasn't done so since.
- Second, the United States should explore with China, Japan, and South Korea the possibility of taking a plutonium production time out. Japan, South Korea, and the United States should offer to delay their plutonium production and fast reactor programs if China would do likewise.

- Third, regarding uranium enrichment, the United State and China should agree to cap their current enrichment capacities and place as much of them under international safeguards for as long as seems reasonable given projected, relatively flat firm commercial demand.

II. Background

China's nuclear weapons program is shrouded in secrecy. Unlike the United States, China has never publicly declared the size of its nuclear weapons stockpile, the number of delivery systems of various types, e.g., ICBMs, SLBMs, bombers, and tactical missiles, or the size of the fissile material inventories, i.e., plutonium and highly enriched uranium (HEU), both in weapons and available for weapons. China has not revealed how it produces tritium for weapons since the closure of its plutonium production reactors. China is not a participant in the New START Treaty. One can make estimates from what is known, as is done here, but it should be a U.S. government policy objective to seek greater transparency regarding China's nuclear weapons-related materials and production capacities.

That said, if we stick to what we know, or what we think we know, we draw the following conclusions:

- China has approximately 350 nuclear warheads, of which roughly 272 are for delivery by more than 240 operational land-based ballistic missiles, 48 sea-based ballistic missiles, and 20 nuclear gravity bombs assigned to bombers.[14]

- China has an estimated 2.9 ± 0.6 t of stockpiled WGPu. This would be enough to make 830 ± 210 nuclear weapons or more, assuming 3.5 ± 0.5 kg WGPu per device.

- China also has stockpiled 14 ± 3 t of weapon-grade HEU in weapons and available for weapons — enough to make 930 ± 370 nuclear warheads assuming 15 ±5 kg of HEU per warhead.[15]

- China can more rapidly expand its HEU production capacity than it can expand its WGPu production capacity. The latter would require China to build more spent reactor fuel reprocessing capacity. It is currently doing so by building two 200 t/y plants. One is expected to come online in 2025, the other sometime before 2030.

---

14. See U.S. Department of Energy, DOE Announces Measures to Prevent China's Illegal Diversion of U.S. Civil Nuclear Technology for Military or Other Unauthorized Purposes," October 11, 2018, available at https://www.energy.gov/articles/ doe-announces-measures-prevent-china-s-illegal-diversion-us-civil-nuclear-technology.
15. Hui Zhang, *China's Fissile Material Production and Stockpile,* IPFM Research Report No. 17, 2018 (Princeton, NJ: Princeton University), available at https://www.belfercenter.org/publication/chinas-fissile-material-production-and-stockpile.

- Another option for doing so is for China to buy the large 800 t/y spent reactor fuel reprocessing plant that it has been in negotiations with France to buy now for several years. This plant would enable China to produce some 1,200 more nuclear weapons' worth of RGPu/y. So far, the French and Chinese have been unable to finalize the deal.

- Thus, to expand its arsenal of two-stage thermonuclear weapons (where the primary would use plutonium), the availability of WGPu is the key material that limits the rate of increase in the size of China's high-yield thermonuclear warhead arsenal ideal for use on its intercontinental ballistic missiles (ICBMs and SLBMs).

- The availability of tritium for boosting thermonuclear warhead primaries is assumed not to be a pacing item, although there is no public information on where and how China currently produces tritium for weapons.

- All of China's dedicated military plutonium production reactors and military chemical separation plants appear to be permanently shut down.

- Should China seek to increase its military WGPu stockpile, it is likely to use its civil reactors to acquire the additional plutonium. In order of priority, China first would reprocess the irradiated blanket material from the two 600 MWe fast breeder reactors it is now building (projected to come on line in 2023 and 2026). If additional WGPu is desired, the next source would be its heavy water reactors, and lastly, it could turn to its light water reactor (LWR) fleet.

- By 2030, China could recover at least 1.2 - 2.0 t of WGPu by reprocessing breeder reactor blanket fuel assemblies, but not touching axial blanket material.[16]
- Used nuclear fuel reprocessing capacity is the pacing item limiting the recovery of WGPu from civil reactors.
- Although there would be a considerable reduction in yield-to-weight and yield-to-volume, single stage fission weapons and even two-stage thermonuclear weapons can be made without WGPu by relying solely on HEU, RGPu, or a combination of both. China is therefore essentially unconstrained in the rate of production of bomber weapons and lower yield ballistic missile warheads.

III. Frequently Asked Questions

Q1. Is China currently pursuing a minimal nuclear deterrence policy?

In the 1990s, China clearly professed a minimal nuclear deterrence policy. China's economy was weaker than it is today. The United States and Russia had comparatively

16. See Table 1.01 below. Totals have been rounded.

far more deployed and reserve nuclear weapons and fissile material for weapons. To make up for this difference, China relied on underground siting of its ICBMs and probably targeted U.S. cities.

Today, China's profession of its minimal deterrence strategy is less clear. It was reported that China has launch-on-warning and launch-on-alert capabilities as well as higher accuracy, lower-yield nuclear missile systems that are suitable for hitting not just area targets (cities), but military point targets.

Q2. How many nuclear weapons do experts believe the Chinese have today? What uncertainties do they assign to their estimates? What is the basis for these uncertainties?

The FAS, a non-government organization that carefully tracks nuclear weapon programs globally, estimates that China "has a produced a stockpile of approximately 350 nuclear warheads, of which roughly 272 are for delivery by more than 240 operational land-based ballistic missiles, 48 sea-based ballistic missiles, and 20 nuclear gravity bombs assigned to bombers. The remaining 78 warheads are intended to arm additional land- and sea-based missiles that are in the process of being fielded."[17]

---

17. Hans M. Kristensen and Mark Korda, "Nuclear notebook: Chinese nuclear forces, 2020," *Bulletin of the Atomic Scientists*, available at

The U.S. Defense Intelligence Agency (DIA) in 2019 estimated that China had a nuclear arsenal in the low 200s, roughly FAS's estimate at the time.[18] It appears from the FAS analyses, historically various U.S. intelligence agencies and elements within the Department of Defense (DOD) have not always been consistent in their estimates of the size of the Chinese nuclear arsenal, and in DIA's earlier projections have not always been consistent with its subsequent estimates. In other words, it appears that the U.S. intelligence community may not have a good handle on the size of the Chinese nuclear arsenal or China's nuclear weapons plans for the future. Nevertheless, the FAS now estimates the current size of the Chinese nuclear arsenal is in the range of about 300 to 350 nuclear warheads.[19]

---

https://thebulletin.org/premium/2020-12/nuclear-notebook-chinese-nuclear-forces-2020/.

18. Hans Kristensen, "DIA Estimates for Chinese Nuclear Warheads," Federation of American Scientists, May 31, 2019, available at https://fas.org/blogs/security/2019/05/chinese-nuclear-stockpile/.

19. See Bunn, Matthew, Hui Zhang, and Li Kang. "The Cost of Reprocessing in China." Project on Managing the Atom, Belfer Center for Science and International Affairs, Harvard Kennedy School, January 2016, available at https://www.hks. harvard.edu/publications/cost-reprocessing-china.

Q3. Is China now attempting to acquire a much larger nuclear arsenal, on par with, or exceeding, that of the United States and Russia?

Since the 1990s, China's economy has grown considerably—to the point that it rivals the size of the U.S. economy. Under the leadership of Xi Jinping, who became president of the People's Republic of China (PRC) in 2013, is the head of the Chinese Communist Party, and paramount leader of China, China's goal is to displace the United States as the most powerful country in the world. To achieve this goal, it may believe it will need a nuclear arsenal as large or larger than that of the United States and Russia, i.e., an arsenal in five to ten years numbering in the thousands, rather than the hundreds of warheads. Its options are to increase the size of its own arsenal or wait until the United States and Russia negotiate a reduction of their nuclear arsenals. To date, China has been unwilling to enter arms control negotiations with the United States and Russia that could place limits on its own nuclear arsenal.

In 2019, the head of the Defense Intelligence Agency (DIA) stated that the Chinese nuclear weapons stockpile will "at least double" over the next decade.[20] A mere

---

20. See National Research Council 2008. Review of DOE's Nuclear Energy Research and Development Program (Washington, DC: The National Academies Press, 2008), pp. 73-76, available at https://www.nap.edu/cart/download.cgi?record_id=11998&file=73-76.

doubling of its stockpile would not put China's nuclear arsenal on par with that of the United States or Russia. A further increase several times more could do so.

Q4. Does China have a breakout strategy to increase the size of its nuclear arsenal? If so, does this strategy include using some of its civilian nuclear facilities to produce fissile material for weapons?

Officials in Beijing continue to insist that China has such a small arsenal, it should not be asked to limit it at this time. On the other hand, commentators in Chinese official newspapers have argued that China needs 1,000 nuclear weapons to deter the United States.[21] China also has explicitly called for a fusion of its civilian and military nuclear efforts. In the United States today, the nuclear weapons program is almost entirely separate from its civil nuclear power program. A notable exception is that tritium for nuclear weapons produced in civil reactors. Chinese nuclear activities are not so clearly segregated. Given the size and orientation of China's civilian nuclear program to plutonium fuel cycles and large uranium enrichment plants, there is increasing

---

21. See Yew Lun Tian, "China needs more nuclear warheads: Global Times editor," *Reuters*, May 8, 2020, available at https://www.reuters.com/article/us-china-usa-war/china-needs-more-nuclear-warheads-global-times-editor-idUSKBN22K1EE.

uncertainty as to what China might do to mobilize and exploit this "peaceful" nuclear infrastructure for military purposes.

Q5. What materials matter most for nuclear weapons?

Were China to substantially increase the size of its nuclear weapon arsenal, China most likely would need to acquire additional stocks of plutonium, HEU, and tritium. Although typical thermonuclear weapon designs contain some five or more times as much HEU relative to plutonium, the pacing item for expanding China's arsenal of two stage thermonuclear warheads would be plutonium, not HEU for two reasons. First, China already has the capacity to make massive amounts of HEU but relatively no confirmed capacity to make WGPu in reactors dedicated for this purpose. Second, China will likely want plutonium to make higher yield-to-weight and higher yield-to-volume thermonuclear weapons for intercontinental missiles.

As for tritium, China has and can get all that it needs. Despite its radioactive decay with a half-life of 12.3 years, tritium can be readily made in military or civil nuclear reactors and should not be a constraint on expanding the weapon arsenal.

Although there would be a reduction in yield-to-weight and yield-to-volume, single stage fission weapons and even two-stage thermonuclear weapons can be made without WGPu by relying solely on HEU, RGPu, or a combination of both. As China modernizes its missiles to carry multiple warheads, it can produce the massive blast effects of a large-yield weapon by laying down much smaller yield multiple warheads. Bottom line: China is essentially unconstrained in the rate of production of bomber weapons, including bombs for strategic bombers and lower yield ballistic missile warheads.

Q6. How much fissile material do experts believe the Chinese have stockpiled for weapons? What uncertainties do they assign to their estimates? What is the basis for these uncertainties? What is the range of surplus weapons plutonium and uranium stockpiled available for use in new weapons?

Non-government best estimates of the Chinese military plutonium stocks have ranged from 1.8 t to 4.35 t with large uncertainties.[22] The best analyses are by Hui Zhang, whose latest estimate is 2.9 ± 0.6 t of WGPu.[23]

---

22. Hui Zhang, "China's Fissile Material Production and Stockpile," available at https://www.belfercenter.org/publication/chinas-fissile-material-production-and-stockpile.
23. Hui Zhang, "China's Uranium Enrichment and Plutonium Recycling 2020-2040: Current Practices and Projected Capacities," paper prepared for the

The uncertainty "mainly come from the assumptions concerning the phase-out of plutonium production during the 1980s.[24]

Assuming 3.5 ± 0.5 kg of WGPu per thermonuclear warhead, this would be sufficient for about 830 ± 210 warheads. Assuming Hui Zhang's analysis is correct, China would have sufficient military plutonium stocks to double its nuclear arsenal over the next decade, recalling, as indicated above, that in 2019 the head of DIA reportedly stated that the Chinese nuclear weapons stockpile will "at least double" by 2030.

Hui Zhang estimates China currently has 14 ± 3 t of weapon-grade HEU available for weapons.[25] Assuming 20 ± 5 kg of HEU per thermonuclear warhead, this would be sufficient for 930 ± 370 thermonuclear warheads. 10 to 15 kg of HEU would likely be used to produce single-stage weapons with yields of a few tens of kilotons and more than 30 kg of HEU would probably be used to

---

Nonproliferation Policy Education Center (NPEC), June 23, 2020, p. 2., available at http://npolicy.org/article_file/Hui_Zhang_Paper.pdf.

24. Hui Zhang, "China's Fissile Material Production and Stockpile," available at https://www.belfercenter.org/publication/chinas-fissile-material-production-and-stockpile.

25. Hui Zhang, "China's Uranium Enrichment and Plutonium Recycling 2020-2040, p. 4, available at http://npolicy.org/article_file/Hui_Zhang_Paper.pdf.

produce weapons with yields in the range of 50 to 100 kilotons.[26]

Q7. How much of the future plutonium, HEU, and tritium production might be produced in facilities that might be viewed as civil facilities?

There are no known military plutonium production reactors operational today. It is therefore likely that beyond existing stocks, any new plutonium for weapons produced in the next decade will come from what are traditionally viewed as civil reactor facilities. There is one caveat here. China has attempted to build and operate an underground plutonium production reactor, so it cannot be ruled out that such a plant exists today (see Q10 below).

Beyond existing stocks, additional HEU for weapons will come from what are traditionally considered to be dual purpose (civil and military) uranium enrichment facilities. There are no treaty agreements, e.g., the Non-Proliferation of Nuclear Weapons Treaty (NPT), or

---

26. Thomas B. Cochran and Christopher E. Paine, "The Amount of Plutonium and Highly Enriched Uranium Needed for Pure Fission Nuclear Weapons," (Washington, DC: Natural Resources Defense Council, Revised 13 April 1995), Figure 2. Such estimates are not an exact science, as the amount of fissile material used would depend not only on the technical sophistication of the warhead design, but also on considerations such as the desired weight and size of the weapon, i.e., yield-to-weight and yield-to volume considerations.

international agreements, e.g., IAEA safeguards agreements, prohibiting the use of civil reactors for the production of tritium for nuclear weapons.

Q8. How much plutonium might China produce by 2030?

The plutonium production reactors at China's two known military plutonium production sites, Jiuquan Atomic Energy Complex (Plant 404) and Guangyuan plutonium-production complex (Plant 821), are believed to have ceased production of plutonium around November 1986 and 1984, respectively.[27] It is reasonable to assume these reactors could not or will not be restarted at this late date. Therefore, it is reasonable to assume that beyond existing stocks, if China desired additional plutonium for weapons prior to 2030, it could turn to its civil reactors for additional plutonium. Estimates below indicate that China could readily produce on the order of 1.2 to 2 t of additional WGPu by 2030 from fast breeder reactors. Additional WGPu (several times as much) could be recovered from civil reactor spent fuel, but this would require substantial increase in nuclear fuel reprocessing capacity.

---

27. Hui Zhang, "China's Uranium Enrichment and Plutonium Recycling 2020-2040, p. 4, available at http://npolicy.org/article_file/Hui_Zhang_Paper.pdf.

Q9. Discuss what challenges China would face to reprocess all of the materials it might want to in a "crash" ramp up scenario. Would it need to build additional reprocessing facilities? What might this entail?

All of China's known chemical separation plants have been shut down for years. China has a pilot reprocessing plant that is believed to be dedicated to reprocessing spent fuel from the Daya Bay Nuclear Power Plant, the site of two 984 MWe LWR. China also is building reprocessing plants with a capacity of 200 tons of heavy metal per year (tHM/y) each. The first is expected to come on line in 2025, the second sometime before 2030. These plants may be fully utilized to separate plutonium to make mixed-oxide (MOX) fuel for the two 600 MWt fast breeder reactors under construction.

China has pursued the purchase of an 800 tHM/y civil reprocessing plant from France. If a plant of this size is built and operated at an average 75 percent capacity factor, it could recover on the order of 6 t/y of plutonium (or 1,200 bombs' worth of RGPu assuming 5 kgs per warhead and that the spent fuel processed contains approximately one percent plutonium).

Q10. What types of military production reactors does China have; what are their years of service and estimated capacity factors? Are there any such facilities now under construction?

Two Chinese military plutonium production reactors, both graphite-moderated water-cooled reactors, have been identified in the unclassified literature. One production reactor is at the Jiuquan Atomic Energy Complex (Plant 404) located 40° 13' 22.76" N; 97° 21' 21.09" E, in Gansu Province.[28] The second reactor (at 32° 29' 44.6" N; 105° 35' 24.5E), is at the Guangyuan nuclear complex (Plant 821) about 24 km WNW of Guangyuan in Sichuan Province.[29]

Hui Zhang has referenced a Chinese press report regarding an unfinished underground Fuling plutonium production complex, the so-called "Plant 816" in Sichuan province. [30] Plant 816 was an effort to build underground

---

28. Robert S. Norris, Andrew S. Burrows and Richard W. Fieldhouse, *nuclear weapons Databook, Volume V, British, French and Chinese Nuclear Weapons* (Boulder; Westview Press, 1994), p. 339.
29. Hui Zhang, "China's Uranium Enrichment and Plutonium Recycling 2020-2040: Current Practices and Projected Capacities," p. 2. Hui Zhang, "China's Fissile Material Production and Stockpile," available at https://www.belfercenter.org/publication/chinas-fissile-material-production-and-stockpile.
30. Hui Zhang, "China's Stockpile of Military Plutonium: A New Estimate," Project on Managing the Atom,

three light-water cooled, graphite-moderated plutonium production reactors each 80 MWt.[31] Construction of Plant 816 was initiated in February 1967 and terminated in 1984.[32]

The lack of knowledge of the actual thermal power ratings of the Jiuquan and Guangyuan reactors, and the lack of assurance that there is not a still secret underground plutonium production facility are the greatest impediments to understanding the size of the inventory of plutonium in and available for Chinese nuclear weapons.

Hui Zhang reports the reactor at Jiuquan operated from December 1966 to 1984.[33] The reactor at Guangyuan started operation near the beginning of 1974. When the Guangyuan reactor ended production is unknown, but it is assumed to have been sometime in the late-1980s or early 1990s.

Kennedy School of Government, Harvard University, 17-21 July 2011.
31. Ibid.
32. Ibid.
33. Ibid.

Q11. What are China's military chemical separation facilities, their capacities, and their years of service? Are there any such facilities now under construction?

China operated three military chemical separation plants all of which are believed to be shut down. These are the Jiuquan intermediate pilot plant (1968 to early-1970s), the Jiuquan chemical separation plant (1970-ca. 1987), and the Guangyuan chemical separation plant (1976-ca. 1987). The capacity of these plants is assumed to have been matched approximately to the rate of plutonium production from the production reactors at these sites.

Q12. What Chinese fissile production-related facilities are not clearly dedicated for military purposes that could be so used?

Plutonium Production Facilities

While there is no public evidence that it intends to do so, China could produce plutonium for weapons by reprocessing spent fuel from civil reactors. Assuming the objective would be to recover WGPu, the order of preference would probably be:

a) fast breeder reactor blanket material;

b) heavy water reactor spent fuel;

c) low-burnup spent fuel discharged from LWRs at the time of the first refueling.

Fast Reactors

The Chinese Experimental Fast Reactor (CEFR) is a 65 MWt, 20 MWe reactor located at (39° 44' 26" N, 116° 1' 51" E), about 20 miles southwest of Beijing. It achieved criticality in July 2010 and was connected to the grid in July 2011. Fuel for the first core consisted of HEU (64.4% enriched) and it seems to still be using HEU in 2020. Later fuel loadings may be MOX fuel: 219.2 kg of fissionable material: --121.6 kg Pu (93.2 kg Pu-239); --97.6 kg U-235 (30% enriched uranium).[34]

Two commercial-size Chinese Fast Reactors (CFRs) are under construction at Xiapu in Fujian province (26° 48' 13" N, 120° 9' 18" E). These CFR-600s are 1500 MWt, 600 MWe reactors. Construction of the CFR-600s began in December 2017 and December 27, 2020, respectively. Commissioning of the first reactor is expected in 2023;

---

34. "China's Fast Breeder Reactor (FBR) Program," Nuclear Threat Initiative, updated February 6, 2004, available at https://web.archive.org/web/20080208125200/http://www.nti.org/db/china/fbrprog.htm. The core has also been reported as containing 150 kg Pu (98 kg Pu-239). *World Nuclear News*, "China starts building second CFR-600 fast reactor," *World Nuclear News*, December 29, 2020, available at https://world-nuclear-news.org/Articles/China-starts-building-second-CFR-600-fast-reactor.

the second in 2026. The maximum burn up is about 100 MWd/kgHM. They will be fueled using HEU supplied by TVEL in Russia.[35] Later they would be fueled by MOX, where the plutonium is recovered at the 200t/yr reprocessing plant. Both the MOX fabrication plant and the reprocessing plant are located at Jinta, Gansu.[36]

Xiapu in Fujian province is also reported to be the site for the prototype so-called travelling-wave reactor TWR-P. In December 2013, a US Federal Register notice said that the USA had negotiated an agreement with China "that would facilitate the joint development of TWR technology" from TerraPower, including standing wave versions of it. In September 2015, CNNC and TerraPower signed an agreement to work towards building a prototype 600 MWe TWR unit in China, apparently over 2018 to 2023.[37] On October 11, 2018, the U.S. Department of Energy effectively killed this project by promulgating a

---

35. "CFR-600," Wikipedia, last modified January 15, 2021, available at https://en.wikipedia.org/wiki/CFR-600#:~:text=The%20CFR%2D600%20is%20a,Xiapu%20fast%20reactor%20pilot%20project.
36. Hui Zhang, "China is speeding up its plutonium recycling programs," *Bulletin of the Atomic Scientists*, Vol.76, No.4, 2020, available at https://www.tandfonline.com/doi/abs/10.1080/00963402.2020.1778372?journalCode=rbul20.
37. "Fast Neutron Reactors," World Nuclear Association, updated December 2020, available at https://www.world-nuclear.org/information-library/current-and-future-generation/fast-neutron-reactors.aspx.

regulation prohibiting the transfer of U.S. advanced reactor technology to China.[38]

A commercial-scale fast breeder, the CFR1000, has been proposed to have a capacity of 1,000-1,200 MWe. An earlier design of the CFR1000 was a 2,500 MWt pool-type reactor using MOX fuel. Subject to a decision to proceed, construction could start in December 2028, with operation from about 2034.[39] That design will use metal fuel and 120-150 GWd/t burnup.

Heavy Water Reactors

Qinshan Phase III, units 1&2, are CANDU-6 pressurized heavy water reactors (PHWR). The Qinshan site is located in Haiyan County ((30° 29' 59" N, 120° 57' 0" E), Zhejiang Province, on the shore of Hangzhou Bay off the East China Sea and approximately 126 km southwest of Shanghai. They are each designed to operate in 2064 MWt, 528 MWe gross, 677 MWe net. They achieved first criticality on September 21, 2002, and January 18, 2003, and were commissioned (began commercial operation) on

---

38. U.S. Department of Energy, "DOE Announces Measures to Prevent China's Illegal Diversion of U.S. Civil Nuclear Technology for Military or Other Unauthorized Purposes," October 1, 2018, available at https://www.energy.gov/articles/doe-announces-measures-prevent-china-s-illegal-diversion-us-civil-nuclear-technology.
39. Hui Zhang, "China's Fissile Material Production and Stockpile," available at https://www.belfercenter.org/publication/ chinas-fissile-material-production-and-stockpile.

December 31, 2002 and July 24, 2003, respectively. The two PHWRs have each operated at high cumulative capacity factor, 0.9, over their lifetime (from the year following criticality through 2019), [40] reflecting the fact that they do not have to shut down for refueling.

The CANDU-6 reactor core has 380 fuel channels contained in and supported by a horizontal cylindrical vessel known as the calandria.[41] There are 12 fuel bundles per channel, each containing 21.4 kg of natural uranium oxide (UO_2) fuel. Typically, in a CANDU-6 about 10 fuel channels per week are refueled, and either four or eight of the fuel bundles in a channel are replaced during a refueling operation.[42]

---

40. "QINSHAN 3-1," International Atomic Energy Agency (IAEA) Power Reactor Information System (PRIS), updated February 16, 2021, available at https://pris.iaea.org/PRIS/CountryStatistics/ReactorDetails.aspx?current=69; and "QINSHAN 3-2," IAEA PRIS, updated February 16, 2021, available at https://pris.iaea.org/PRIS/CountryStatistics/ReactorDetails.aspx?current=840.
41. Zhang Yanfa and B.A. Shalaby, "Introduction to the Qinshan Phase III CANDU Nuclear Power Plant," CANTEACH, 1999, available at https://canteach.candu.org/Content%20Library/20054402.pdf.
42. *CANDU 6 Technical Summary*, CANTEACH, May 2005, available at https://canteach.candu.org/Content%20Library/CANDU6_TechnicalSummary-s.pdf.

Light Water Reactors

China had 49 operational reactors in 2020, of which 46 were LWRs.[43] Many of the LWRs were designed by foreign nuclear vendors and all contain some technology from foreign vendors. China has been moving toward a goal of supplying 100 percent of the nuclear components from domestic suppliers.

Q13. What amount of WGPu is currently available from these sources and their projected annual production? Identify similar sources from reactors under construction. What international agreements, e.g., with the International Atomic Energy Agency (IAEA) or nuclear supplier nations, are intended to prevent the use of civil reactors or civil nuclear fuel reprocessing plants for military purposes?

Fast Breeder Reactors

The CEFR is 65 MWt and the CFR-600s are 1500 MWt each. Here, as other analysts have done, we will scale from a WGPu production estimate for the PFBR, a 1250 MWt sodium-cooled fast breeder reactor in India. Alexander Glaser and M.V. Ramana estimate that the

43. "Country Statistics for People's Republic of China," IAEA PRIS, updated February 16, 2021, available at https://pris.iaea.org/PRIS/CountryStatistics/CountryDetails.aspx?current=CN.

PFBR, using MOX fuel in the core and operation at a capacity factor of 0.75, will produce 92.4 kg WGPu/y in the radial blanket and 52.0 kg WGPu/y in the axial blankets.[44] Scaling by (1500/1250 =) 1.2, the CFR-600 operating at the same capacity factor of 0.75 and using MOX fuel in the core will produce an estimated 173 kg WGPu/y in the radial and axial blankets combined, of which 111 kg WGPu/y are produced in the radial blanket.

The first core of the first CFR-600, and possibly the first core of the second CFR-600, will be fueled with enriched UO_2, rather than MOX. In Table 1.01 we therefore reduce the rate of production for the first three years by five percent. The low end of the range assumes the reactors operate at a capacity factor of 0.65 and only the plutonium from the CFR-600 radial blankets is recovered for the weapons stockpile, and the upper limit assumes a capacity factor of 0.75 and the WGPu from the radial and both axial blankets is recovered. We assume that the MOX cores of these reactors were fueled with RGPu recovered from reprocessed LWR spent fuel, and the WGPu bred in the reactor blankets is not recycled but could be reserved for weapons.

---

44. Alexander Glaser and M. V Ramana, "Weapon-Grade Plutonium Production Potential in the Indian Prototype Fast Breeder Reactor," *Science and Global Security*, 15:85-105, 2007, Table 2, p. 94, available at https://www.tandfonline.com/doi/abs/10.1080/08929880701609154.

Table 1.01: WGPu recovered from the blanket material in Chinese fast breeder reactors. Totals should be rounded.

Year	**CEFR** (kg WGPu)	**CFR-600s)** (kg WGPu)	**Cumulative** (kg WGPu)	**Annual Total** (kg WGPu)
2012-2020	45 - 56			
			45 - 56	
2021	5 – 7			5 – 7
			50 - 63	
2022	5 – 7			5 – 7
			55 - 70	
2023	5 – 7			5 – 7
			60 - 77	
2024	5 – 7	91 - 164		96 - 171
			156 - 248	
2025	5 – 7	91 - 164		96 - 171
			252 - 419	
2026	5 – 7	91 - 164		96 - 171
			348 - 590	
2027	5 – 7	187 - 337		192 - 344
			540 - 934	
2028	5 – 7	187 - 337		192 - 344
			732 - 1,278	
2029	5 – 7	187 - 337		192 - 344
			924 - 1,622	
2030	5 – 7	192 - 346		197 - 353
			1,121 - 1,975	
Total	**95 - 126**	**1,026 - 1,849**		

Heavy Water Reactors

The heavy water reactors in China are Canadian designed CANDU reactors under IAEA safeguards. Therefore, the used fuel cannot be reprocessed to recover plutonium for weapons without violation of China's agreement with the IAEA and Canada. Also, China currently does not have adequate nuclear fuel reprocessing capacity to process the large amount of spent CANDU fuel to obtain significant amounts of WGPu. Thus, while the following calculations show how much WGPu could be produced in theory, it is unlikely that China will seek WGPu from its CANDU reactors.

Under normal operating conditions, a CANDU reactor is a relatively sparse generator of RGPu as a weight percent of its used fuel. To obtain 800 kg of RGPu (at 27 percent Pu-240) over 200 tons of irradiated CANDU fuel (over 10,000 spent fuel bundles) would have to be reprocessed.[45] The civil demonstration reprocessing plant under construction near Jinta in Gansu province, has a rated capacity of 200 t/y.

A typical fuel burnup attained in the CANDU-6 is 7,500 MWt-d/tHM.[46] If fueled with natural uranium (Nat-U), the spent fuel contains approximately 0.5 percent RGPu, or about one-half the concentration of RGPu in LWR spent fuel. To produce WGPu in the CANDU it would have to operate at about 1,000 MWt-d/tHM, in which case

---

45. J.J. Whitlock and A.G. Lee, "CANDU: Setting the Standard for Proliferation Resistance of Generation III and III+ Reactors," IAEA-CN-164-5P07, available at https://www-pub.iaea.org/MTCD/Publications/PDF/P1500_CD_Web/htm/pdf/poster/5P07_J.%20Whitlock.pdf.
46. Ben Rouben, "CANDU Fuel-Management Course," CANTEACH, March 1999, available at https://canteach.candu.org/Content%20Library/20031101.pdf.

the concentration of WGPu in the spent fuel would be 7.5 times lower, or about 0.067 percent. Thus, reprocessing such low-burnup spent fuel at the 200 t/y demonstration reprocessing plant under construction near Jinta in Gansu province, operating at an assumed capacity factor of 75 percent, on the order of 100 kg of WGPu could be recovered per year.

The Qinshan Phase III, Units 1 & 2, CANDU-6 reactors, which are each rated at 2064 MWt, are assumed to continue to operate at their historical cumulative capacity factor of 90 percent. Thus, each could produce about 540 kg of WGPu per year. Assuming they were both used to produce WGPu beginning in 2023, then by the end of 2030, together they could produce approximately 8,670 kg of WGPu. When added to the WGPu from the two fast reactors the total estimated WGPu production through 2030 would be 9,950 kg to 10,300 kg. However, this would require far more spent fuel reprocessing capacity than China has contemplated.

Alternatively, if the two CANDU-6 reactors operated at a burnup of 2,000 MWt-d/tHM for nine months, they could produce 1,018 kg of FGPu containing 9.8 percent Pu-240. This could be blended with 1,278 kg of super-grade Pu, the lower end of the estimated WGPu produced in the two fast reactors by 2030, to make about 2,300 kg of 6 percent Pu-240 WGPu. However, since China lacks reprocessing capacity, it is most likely that China would not produce super-grade Pu in the fast reactor blankets. Rather China would leave the radial blankets in the reactor for a period twice as long, thereby doubling the plutonium concentration and recovering WGPu (something close to 6 percent Pu-240).

Clearly, the bottleneck in producing additional WGPu by processing CANDU fuel is the lack of spent fuel reprocessing capacity. Also, there may be restrictions imposed by IAEA safeguards agreements or China-Canadian bilateral agreement(s) covering the CANDU reactors.

Gregory Jones has made a similar estimate of the amount of WGPu (at 6 percent Pu-240) that could be produced when operating at 1,200 MWt-d/tHM, compared to operating at 7,000 MWt-d/tHM.[47] Jones concludes:

> . . .this could not be done for the entire reactor, since it would require using 5.8 times as much fuel per year and the reactor's refueling machine could not operate quickly enough. If this rapid refueling were limited to only one-eighth of the core, it would be feasible. Each reactor would then produce 67 kg WGPu per year. The plutonium would be at a concentration of only 1 kilogram per metric ton, so reprocessing the low burnup fuel from both reactors would require 134

---

47. Gregory S. Jones, "China's Current Stocks of Separated Plutonium and Possible Growth to 2040," September 10, 2020.

> metric tons per year of reprocessing capacity. This would be a significant fraction of China's reprocessing capacity. Finally, as was noted by von Hippel and Takubo, China's peaceful use agreement with Canada for these reactors requires Canada's permission to carry out the reprocessing, which may deter China from taking this action."[48]

Light Water Reactors

A report by Lawrence Livermore National Laboratory (LLNL) and Stanford University, "Verifying the Agreed Framework," April 2001, devotes a chapter to how the Democratic People's Republic of Korea (DPRK) could exploit LWRs to obtain plutonium for weapons.[49] One of the scenarios for obtaining FGPu from LWRs involves "short cycling," that is operating the reactor at a reduced burnup:[50]

> In the event of overt short-cycling of the reactor with a goal to obtain plutonium with 90 percent 239Pu, burnup would be

---

48. Ibid.
49. Michael May, Ed., "Verifying the Agreed Framework," LLNL, UCRIL-ID-142036 and CGSR-2001-001, April 2001, Chapter 5, available at https://cgsr.llnl.gov/content/assets/docs/CGSR-2001-001.pdf.
50. Ibid., p. 63.

> limited to no more than about 7 Wd/kg, limiting the reactor cycle time to approximately 9 months. Assuming an industry average of 40 days per refueling outage and refueling of the entire core, the reactor could produce as much as 150 kilograms of plutonium (containing approximately 90 percent 239Pu) every 10 months. †
>
> ---
>
> † These data are obtained from the rough approximations of plutonium production in LWRs described earlier. Plutonium production rates vary broadly with fuel enrichment and using lower fuel enrichments would likely reduce cost and optimize plutonium production. Significant additional analysis is required to determine more accurate estimates of fuel requirements and of material production and plutonium isotopic contents at these lower burnups.

China could devote one or a few of its civil LWRs to obtain one to a few hundred kg of FGPu/year, however, this is unlikely to be China's first choice as it would involve operating reactors meant for electricity production inefficiently, and the plutonium is 90 percent Pu-240, not WGPu. To obtain 6 percent or less, Pu-240 would entail operation at a burnup closer to 3.5 MWt-d/kgHM, all but abandoning the use of the reactor for electricity production. As with recovering WGPu from CANDU spent fuel, the bottleneck here is the lack of spent fuel reprocessing capacity.

Q14. Identify the normally civil spent fuel reprocessing plants, their capacity, and years of service. Are there any such facilities under construction?

Since 1983, however, China has maintained a closed-fuel-cycle policy. In December 2010, it began testing a pilot civilian reprocessing plant with a capacity of 50 tons of spent fuel per year at the Jiuquan complex. Since then, however, the plant has been shut down most of the time because of technical problems. As of December 2015, according to China's declaration to the IAEA, it had separated 25.4 kilograms of plutonium. In early 2015, the government approved the construction of a 200 ton/year demonstration reprocessing plant at Jinta in Gansu province; CNNC started site preparation for the project that July (40° 19' 29.7" N; 98° 30' 53.3" E) and it is expected to be commissioned in 2025.[51] China recently began construction of another identical plant that is expected to come on line sometime before 2030.[52] CNNC also has been negotiating with France's Areva over the

---

51. For details see Zhang, Hui. "Pinpointing China's New Plutonium Reprocessing Plant," *Bulletin of the Atomic Scientists,* May 5, 2020, available at https://thebulletin.org/2020/05/pinpointing-chinas-new-plutonium-reprocessing-plant/.
52. Hui Zhang, See Hui Zhang, "China starts construction," available at http://fissilematerials.org/blog/2021/03/china_starts_construction.html.

purchase of a commercial reprocessing plant with a capacity to reprocess 800 tons of spent fuel per year.[53]

Q15. Identify the nominally civil uranium enrichment facilities, their types, years of service, and estimated capacity.

Typically, HEU for military and civil use are produced at the same facilities. As noted previously, HEU is not a pacing material for China's nuclear weapons production. China has three large operational gas centrifuge uranium enrichment plants (Lanzhou, Hanshong, and Emeishan) with a combined capacity estimated to be on the order of 7.8 million SWU/year.[54]

The Lanzhou gas centrifuge enrichment complex (Plant 504) alongside the Yellow River near Lanzhou in Gansu province (36°08'53.30'' N; 103°31'24.49'' E) has an estimated capacity of 2.6 million SWU/year.[55]

---

53. Hui Zhang, "China's Fissile Material Production and Stockpile," IPFM p. 20, available at https://www.belfercenter.org/publication/chinas-fissile-material-production-and-stockpile.
54. Hui Zhang, "China's Uranium Enrichment and Plutonium Recycling 2020-2040," p. 2, available at http://npolicy.org/article_file/Hui_Zhang_Paper.pdf.
55. Ibid., p. 3 and Hui Zhang, "China's Fissile Material Production and Stockpile," IPFM, pp. 7-8, https://www.belfercenter.org/publication/chinas-fissile-material-production-and-stockpile.

The Hanzhong enrichment plant (official Chinese name: CNNC Shaanxi Uranium Enrichment Co., Ltd., or Plant 405) has an estimated capacity of 2.7 million SWU/year. It is located at (33°15'47.9" N/107°25'49.5" E), about 45 km NE of Hanzhong in Shaanxi province. It was built in three stages with the first two stages using Russian made centrifuges.

The Emeishan enrichment plant (Plant 814) has an estimated capacity of 2.45 million SWU/year. It consists of a pilot plant (0.25 million SWU/year) at (29°38'38.9 N/103°29'23" E) and a larger plant (2.2 million SWU/year) at (29°40'42 N/103°32'3.6" E), mid-way between Emeishan and Jiajiang, and about 5.75 km northeast of the pilot plant.

Two older gaseous diffusion uranium enrichment plants, at Lanzhau and Heping are now shut down.[56] HEU production in China began at Lanzhau on January 14, 1964.[57]

Besides the centrifuge facilities at Lanzhou, Hanzhong, and Emeishan, CNNC also plans to build two larger uranium-processing complexes in Hebei (referred to as "North Project") and Guangdong (referred to as "South Project").[58] China exports and imports low-enriched

56. Ibid., p. 6.
57. Ibid., p. 9.
58. Hui Zhang, "China's Uranium Enrichment and Plutonium Recycling 2020-2040, p. 8, available at http://npolicy.org/article_file/Hui_Zhang_Paper.pdf, cites

nuclear fuel, in effect exporting and importing the SWU needed to enrich the fuel.

Addendum
Technical Issues Related to Nuclear Weapons Production

A. Plutonium Production in Reactors

Plutonium-239 is produced in reactors through the absorption of neutrons in U-238 followed by two subsequent beta decays:

$$n + U^{238} \longrightarrow U^{239} \xrightarrow[24m]{\beta-} Np^{239} \xrightarrow[2.4d]{\beta^-} Pu^{239}$$

Production of plutonium in quantity requires copious supply of neutrons. This is achieved in a reactor, where a sustained and controlled chain reaction of fissioning U-235 nuclei provides a steady flux of neutrons to bombard U-238. The rate of production of plutonium depends on the design of the reactor and its fuel and the total energy production of the reactor, i.e., the operating power of the reactor integrated over time. If the reactor is designed to operate at a low power level, the neutron flux will be small and plutonium production will be low. Also, if the reactor is fueled with HEU, and there is no additional U-238 target material present, then the absence of U-238 in quantity results in very little plutonium production. This is the case

---

CNNC stepped up deployment of nuclear fuel industrial park to help China 's nuclear power program go global. See *China Daily*, January 29, 2016, available at http://www.chinadaily.com.cn/interface/yidian/1120781/2016-01-29/cd_23312141.html.

with U.S. naval reactors fueled with 97.3 percent-enriched HEU.

In addition to Pu-239, heavier isotopes of plutonium, namely, Pu-240, Pu-241, and Pu-242 are also made by subsequent neutron absorption:

$$n + Pu^{239} \longrightarrow Pu^{240}$$

$$n + Pu^{240} \longrightarrow Pu^{241}$$

$$n + Pu^{241} \longrightarrow Pu^{242}$$

Also, Pu-238 is created by two sources: neutron capture by Np^{237} followed by beta decay of Np^{238}; and alpha decay of Cm^{242} which comes from the beta decay of Am^{242}, which in turn comes from neutron capture of Am^{241}.

The rate at which these competing reactions take place in the reactor, and the rate at which the various plutonium isotopes build up, are dependent on the design of the reactor and its fuel, and on the rate of energy production of the reactor. The plutonium is characterized by the percentage of Pu-240 it contains:

Grade of Plutonium	**Percentage of Pu-240**
Super-grade	2-3%
Weapon-grade (WGPu)	less than 7%
Fuel-grade (FGPu)	7% to less than 19%
Reactor-grade (RGPu)	19% or greater

For a given reactor type, it is useful to express the plutonium isotopic concentration as a function of the fuel exposure, or "fuel burnup." "**Fuel burnup**" can be measured as a percent of the fuel that is fissioned, but here it is more useful to express it as the thermal energy generated per unit of fresh fuel. In this case it can be expressed in units of megawatt-days of thermal power (MWt-d) per metric ton (t) of heavy metal (MWt-d/tHM) or per kilogram (kg) of heavy metal (MWt-d/kgHM). Heavy metal is the mass of the fuel, i.e., uranium and/or plutonium, in the fresh fuel, exclusive of the mass of the fuel cladding material.

a) **Plutonium Production Reactors**. Table 1A displays the plutonium production characteristics of the B-Reactor at Hanford in the United States (now shutdown), a typical graphite-moderated plutonium production reactor

Table 1A. Plutonium production in grams (g) and the percent concentration of plutonium as a function of fuel exposure (mass % as a function of MWt-d/tHM) for the Hanford B reactor.

Fuel Burnup Pu Production Isotopic Concentration (%)

MWt-d/tHM	**g Pu/MWt-d**	**Pu-239**	**Pu-240**	**Pu-241**	**Pu-242**
0	0.96	100	0.0	0.0	0.0
100	0.938	98.99	0.9985	0.01121	4.907×10^{-5}
200	0.917	98.00	1.955	0.04430	0.0003142
300	0.897	97.03	2.874	0.09309	0.001018
400	0.879	96.08	3.756	0.1593	0.002321
500	0.862	95.15	4.605	0.2398	0.004336
600	0.846	94.23	5.42	0.3329	0.007274
673	**0.835**	**93.57**	**6.00**	**0.4162**	**0.01012**
700	0.831	93.24	6.21	0.4373	0.01115
800	0.817	92.46	6.97	0.5517	0.01608
900	0.804	91.51	7.703	0.6748	0.02215
1,000	0.792	90.75	8.411	0.8057	0.02941

As can be seen from Table 1A, to ensure that the plutonium is weapon-grade, the average fuel burnup must be kept less than about 800 MWt-d/tHM. To prevent the buildup of Pu-240 beyond weapon-grade, the uranium fuel must be removed from the reactor early and often, typically after residing in the reactor for only a couple of months or so. Thus, reactors primarily for WGPu production are designed for rapid refueling and require a large amount of fuel to be "pushed" through the reactor.

From the second column of Table 1A, note that as the fuel burnup approaches zero the reactor produces about 0.96 g of Pu-239 per MWt-d of energy production. As the fuel burnup increases some of the plutonium is fissioned. At about 500-600 MWt-d/tHM, typical fuel burnups used for WGPu production in this type of reactor, the reactor produces about 0.85 g of WGPu per MWt-d of energy production.

About the Author

Henry Sokolski is executive director of the Nonproliferation Policy Education Center. He served in the Pentagon as Deputy for Nonproliferation Policy, as a consultant to the National Intelligence Council, a member of the Central Intelligence Agency's Senior Advisory Group, a Senate military legislative aide to a member of the Senate Armed Services Committee, and as a special assistant on nuclear energy to the chairman of Senate's Tennessee Valley Authority oversight committee. Congress appointed him to two congressional commissions on the prevention of WMD proliferation. He has authored and edited numerous volumes on strategic weapons proliferation issues, including *Best of Intentions: America's Campaign against Strategic Weapons Proliferation* and *Underestimated: Our Not So Peaceful Nuclear Future.*

Made in United States
North Haven, CT
25 January 2025

64895061R00065